Praise For All-Electric America

"All-Electric America is a timely and engaging educational book about America's energy future. It rightly focuses on the cleanest and most efficient types of energy and transportation, including relevant discussions of the grid and energy efficient buildings. It amplifies the vision that an electric America powered by clean, renewable energy will produce significant health, environmental, and cost benefits with little downside. I highly recommend this book for lay audiences, students, and those in the energy sector who would like to be kept up to date in this rapidly-changing field." – **MARK Z. JACOBSON, Professor of Civil and Environmental Engineering at Stanford University and director of its Atmosphere and Energy Program.**

"The U.S. has some big problems that require bold solutions. Unfortunately, books about solutions to our society's problems are often given short shrift by reviewers or languish on our bookshelves...Now comes S. David Freeman... [All-Electric America] is scathing but optimistic, and manages to be bold while remaining pragmatic. Drawing on their combined years of experience, Freeman and Parks make the case for addressing the dangers of climate change with some concrete steps to counter our current downward spiral... you will have a chance to make yourself knowledgeable about the real avenues available to us to transform our energy infrastructure for present and future generations by moving toward a new renewable energy economy." – **RALPH NADER, Consumer advocate, lawyer and author.**

"One cannot read this book without coming away with an optimistic belief that we, as humans, can handle this problem. Curing cancer and ending world hunger may be beyond us, but converting to an all-electric future can be done. We have the science and we have the policies, we just need the political will and the perseverance." – **R.F. HEMPHILL, Former Executive VP at AES & CEO of AES Solar Power Ltd.**

"... David Freeman and Leah Parks have written a book whose importance cannot be understated and whose timeliness is critical... it is an inspirational book in that the authors show clearly not only why energy sustainability has to be the most important policy issue of our time, but ways to achieve it." – **DR. DAVID DUKE, Associate Professor & Head of Environmental and Sustainability Studies, Acadia University, Canada.**

"This is an excellent book. I strongly recommend reading it..." – **KARL-FRIEDRICH LENZ, Professor of German Law, European Law and International Trade Law, Aoyama Gakuin University, Tokyo.**

"This book was a surprise and a revelation. What it does, no less, is propose in the strongest terms that America's electric utilities do two things: (1) step forward boldly and supplant energy suppliers that fuel our buildings and transportation sectors, both of them high-energy use and high carbon emitting and (2) as quickly as possible, end use of fossil-fueled generation and replace it with renewable energy...Check it out... the strong vision of the role that clean electricity could play in sectors it has barely touched—buildings and transportation—is compelling." – **ROBERT MARRITZ, Publisher and executive editor, ElectricityPolicy.com and Electricity Daily.**

"All Electric America is an important piece of work that all of us should read. But, it's also a potent reminder of David Freeman's commitment to be in it for the long haul. With much work to do to overcome powerful interests intent on slowing progress; we could all do well to follow his example." – **ROB SARGENT, Energy Program Director at Environment America.**

ALL-ELECTRIC
AMERICA

ALL-ELECTRIC
AMERICA

A Climate Solution and the Hopeful Future

S. David Freeman & Leah Y Parks

SOLAR
FLARE
PRESS

All-Electric America: A Climate Solution and the Hopeful Future
S. David Freeman & Leah Y Parks

S☉LAR
FLARE
PRESS

Solar Flare Press
ISBN 978-0-9961747-2-5
January 2016

Books may be purchased in quantity for educational, business, or promotional use. Authors are available to attend live events.

For information please contact:
info@allelectricamerica.com

Cover Design by: Vraciu Andreea
Editing by: Emily Loose Literary
Copy Editing by: Mary Altbaum
Indexing by: Sanjiv Kumar Sinha
Interior Book Design, and Cover Editing & Lettering by: Gwyn Snider
Interior Book Design and eBook Design by: Yannis Schoinas
Website Design & Social Media by: Michalis Schoinas

www.allelectricamerica.com

ISBN 978-0-9961747-3-2, e-book (e-pub) Edition
ISBN 978-0-9961747-4-9, e-book (MOBI) Edition

Library of Congress Control Number: 2015915852

10 9 8 7 6 5 4 3 2 1

We dedicate this book to

our children and Dave's grandchildren and great-grandchild
Charlie,
William, Eirini,
Carolyn, Kelsey, Tess, Tim, Ben, Alexander, Karen, Nate, Lisa,
Roger, Stanley, and Anita,
and
the children of planet Earth.

About the Authors

The authors, Dave, with over forty years as an energy policy maker and utility CEO, and Leah, a journalist in the electricity industry, formed a partnership because they agree that the time has come for a well-defined twenty-first century solution to our climate change problems. Despite the difference of almost half a century in life experience, they quickly discovered that they had a shared vision of how our energy infrastructure needs to transform. Drawing upon what they have learned from their experiences with the electricity industry and concerned about the lack of any comprehensive strategy by world leaders and many environmentalists to meet the challenges of climate change, they developed the plan of action presented in this book. They provide a clear plan to reach the U.S. and world climate goals and a better All-Electric renewable energy future.

S. David Freeman was the first person in the U.S. government responsible for energy policy back in 1968. He was one of the architects of the EPA during the Nixon administration. He was appointed chairman of the Tennessee Valley Authority by President Jimmy Carter and after seven years at TVA served as CEO of major public utilities for thirty years including: New York Power Authority, Los Angeles Department of Water and Power (LADWP), Sacramento Municipal Utility District, and Lower Colorado River Authority. In 2005 Mayor Antonio Villaraigosa appointed Freeman as the president of the Los Angeles Board of Harbor Commissioners. The board implemented the most aggressive Clean Air Action Plan in the nation. He most recently served as LA's Deputy Mayor for Energy and the Environment, and briefly as interim manager of LADWP until April of 2010.

Mr. Freeman is recognized as an eco-pioneer for advancing energy efficiency and renewable energy and cleaner air for the past forty-five years. He has written and lectured extensively on energy and the environment and is the lead author of the influential report "A Time to Choose" written in 1974, under the auspices of the Ford Foundation that first documented energy efficiency as a major part of energy policy. He has also authored *Energy: The New Era*, written in 1974, and *Winning Our Energy Independence, an Energy Insider Shows How*, written in 2007.

Freeman was featured in the critically acclaimed documentary *Who Killed the Electric Car?* in 2006. Freeman has won awards from the Los Angeles Coalition for Clean Air, National Wildlife Association, Global Green, CEERT, CalStart 2007 Blue Sky Award, and many other organizations for his devotion to clean air and renewable energy.

Leah Y. Parks is an associate editor for ElectricityPolicy.com and Electricity Daily, a journal and newsletter that examine current events and the state of the electricity industry for utility executives, commissioners, regulators, and other experts in the industry. She has carried out extensive research in the energy field, has been on the advisory committee of Smart Grid Northwest, serves as an advisor for Oregonians for Renewable Energy Progress, has acted as an advisor for technology reports, and has written extensively about innovations in energy storage, smart grid technology, and renewable energy.

Ms. Parks holds a Masters of Science degree from Stanford University in Civil and Environmental Engineering and a BA with Honors from the University of Wisconsin in International Relations. Her professional experience includes work at the civil engineering firm CH2M Hill on projects focusing on water distribution, water planning, and resource allocation. Parks's unique and diverse background encompasses expertise in the technical fields of environmental engineering and science as well as the fields of journalism, international relations, biological sciences, languages, public relations, and fine arts.

TABLE OF CONTENTS

PART II: OBSTACLES

PART IV: LET'S MAKE IT HAPPEN

Part I
The Promise

CHAPTER 1

An All-Electric Renewable America

When Thomas Edison invented the incandescent light bulb to replace the kerosene lantern, the technology was clearly superior; the light bulb was much safer and produced better light. Even so, many doubted the viability of electric lighting, challenging the electric companies' claims that the cost of electricity would be about the same, or perhaps a bit more, than that of kerosene. The debate went on for years, and a decade after the light bulb was introduced, it still had not made its way into homes.[1]

In our time a host of modern-day Edisons have invented technology that empowers us to make the urgently needed shift from burning fossil fuels to using only clean, renewable energy. We are at a tipping point moment when renewable energy innovations that have been in development for many years have been improved so much that they are viable for mass adoption and can provide us with all the energy we need. They will also be able to do so at decreasing costs.

We can now power all of our energy needs with electricity generated completely by renewable energy. As with the advent of personal computers, when it seemed we woke up one morning and found computers on every desk and smart phones in everyone's hands, we are poised to wake up to a transformed world in which we are living better electrically. But for that to happen soon enough—in time for us to prevent runaway global warming—consumers, public officials, and our energy companies must wake up to the opportunity.

An all-renewable energy supply will be less expensive and prices will be more stable, free of market manipulation and shocks due to conflict abroad. The U.S. will finally have a robust system of inexpensive "home-grown" fuel sources and will be energy independent.

This vision may come across as a utopian fantasy. After all, despite calls for urgent action on the climate problem for well over a decade, emissions are still growing. The building of our renewable energy infrastructure is following the path of the adoption of the light bulb. Electric cars are now commercial, but they have hardly taken the country by storm. And solar panels still adorn only occasional rooftops in most communities. But we are poised for a twenty-first century energy revolution, and in this book, we will show how possible it is to achieve this renewable-fuel-only future. We'll also show how vital doing so is, not only in order to halt global warming, but because of the great benefits every American household stands to take advantage of.

Much of the argument for moving to renewable energy has been premised on the need to combat the climate challenge, and we agree that doing so is our most urgent mission. But the exciting truth that has gone missing in the debate is the fact that an entirely renewable energy supply will be CHEAPER, more sustainable, and more price stable, as well as cleaner and safer than our existing sources of energy.

Most people believe that we continue to fuel our civilization with poisonous power sources because they are cheaper than renewables. Renewable energy is also characterized as only part-time, an unreliable source of electricity that's also too small in scale to truly replace fossil fuels. The truth is just the opposite. Renewables are cheaper, the sources are superabundant, and it is entirely feasible to produce and distribute electricity generated by them in massive quantities. In fact, the technology is now available to generate enough electricity to meet all of our energy needs many times over by solar and wind sources alone.

We can build a truly clean and sustainable electric system that can fuel all our transportation, heat all of our buildings, and fuel all industrial processes within the next thirty-five years. Transforming the United States' current electricity capacity, which is roughly one million megawatts, will require developing approximately 60,000 megawatts of new renewable capacity a year over a thirty-five-year timeframe to supply all of our energy needs in 2050.[2] In 2015 China planned to install roughly 40,000 megawatts of new renewable power and by 2030 is committed to add a total of 800,000 to one million megawatts of greenhouse gas free energy, the equivalent of today's total U.S. electricity capacity.[3] There is no doubt that America can do the same if we decide to.

Transforming our entire energy infrastructure to run on renewable energy by the year 2050 will require a larger effort than solely changing out our current electricity capacity. Investments in coal mining, oil and gas drilling, and building new large coal, gas, and nuclear plants will give way to a massive increase in the construction of solar and wind power plants. The production of electricity will steadily replace the consumption of fossil fuels. Investments will shift—one year at a time—but the total investment in the energy sector will not increase dramatically. The new technologies—solar and wind—will, over three or four decades, replace coal and petroleum just as the cell phone has replaced the old telephone.

Much has been said in recent years about renewables being unreliable. A reasonable person might ask, "What do we do when the sun is down or the wind doesn't blow? We need electricity all the time, so how can all of our electricity come from solar and wind power?" The answer is STORAGE of renewable generated power for when we need it. We can orchestrate the use of solar, wind, geothermal, and hydroelectric energy, with storage systems and smart technology, to provide reliable power day and night. We already have all of the technology and tools to create a reliable system, and they will keep improving.[4]

Many technologies will be essential components of the transformed energy infrastructure, such as the use of vastly improved heat pumps to tap the enormous resource of heat in the atmosphere and below the Earth's surface, as well as both electric and hydrogen fuel-cell cars, trucks, and trains. But the ensemble of solar and wind power and storage, harnessed in an elegantly balanced system, is the heart of our energy future. This trio could supply all of our needs. When the sun does not shine, the wind often blows, and when neither is generating new power, storage can supply it. Cost-effective technology exists today to achieve this, and there is land aplenty.[5]

Of course the infrastructure will take some years to build. So let's jump ahead to 2050, the time by which a robust renewable-only supply system is achievable, if development is stepped-up now and continues every year. Good estimates are that, with a completely renewable energy infrastructure run on electricity, in that year the total energy demand of the U.S. will be roughly 1.6 terawatts (TW) and the total electric and storage capacity needed will be roughly 6.4 TW.[6] The total new land footprint is estimated to be ~0.42% of the U.S.[7]

We Live in a Solar World

Fossil fuel and oil advocates have portrayed solar power as requiring so many photovoltaic (PV) panels that land requirements make it impractical. They are wrong. The U.S. has more than 150 TW of potential rural solar PV capacity alone on developable land that could be tapped by that time—roughly eighteen times the requirement—and that is excluding areas that should not be developed, such as locations involving critical environmental concern, and federally protected lands.[8]

There is land aplenty in part because of the remarkable potential of our deserts. Our pristine desert land must be preserved, but there are vast expanses of already disturbed land in our deserts that would not be further harmed by installing solar equipment on them. They are our solar gold mines, where the sun beats twelve hours per day, three hundred sixty-five days per year, with six to seven hours of that being direct sunlight. They produce the lowest-cost sun power, in fact at a cost that is lower than that of fossil fuels when amortized over the life of the solar plant.

Even if we focused only on off-ground space in which to install solar panels, such as residential and commercial roofs and building sides, parking lots, along highways, and already disturbed vacant land outside the deserts, we would need to make use of only a small amount of it. According to the National Renewable Energy Laboratory, we have about 140 million acres of this space, and only 7% of that area would be required to meet all of our current electricity needs.[9]

In fact, if you compare solar with coal, we disturb less land in the long run because of one simple fact. To feed a coal-fired power plant with strip-mined coal, you have to strip more and more land every year, which wrenches trees from the ground, devours topsoil, and buries fresh water springs and streams under tons of rubble. With solar, you use the same land year after year, only replacing panels, which last thirty years or more.

Huge Wind Resources

Our wind resources are also bountiful. The U.S. could generate all of its projected 2050 energy needs with wind power alone.[10] And though the cost of wind generation was once prohibitive, we can now realistically view wind as an essential partner to the sun. Not only does it often blow when the sun does not shine, but it is one of the cheapest energy sources today. In the last twenty years, the cost to produce electricity from wind has dropped 85%, with 43% of that being since 2009.[11] Good wind can produce a kilowatt hour (kWh) of electricity at about the same cost or lower than existing natural gas, nuclear, and coal, and building more wind capacity will in fact cost less than building new coal, nuclear, and gas plants.[12] The National Resource Defense Council has found that in the near term the cost is expected to continue to fall and to be competitive with natural gas and coal even in low wind areas.[13]

We must factor in the opposition from people who don't like the looks of wind turbines as well as complaints about noise and harm to wildlife. The "not in my line of sight" or "not in my backyard" argument, commonly known as NIMBY, has succeeded in killing many projects, and many of the environmentally concerned who might support wind generation oppose it due to the perceived threat to wildlife. These concerns are overblown but must not be dismissed; we can show great consideration for such concerns when siting wind farms. Fortunately, the DOE has determined that we can find more than enough land and coastlines where installing wind turbines would not cause serious problems.[14]

A Comparative Bargain

The 2015 drop in the price of oil has been big news, but the more significant news is that the price of solar and wind power has decreased by a far greater percentage. The long-term difference is crucial; both solar and wind power will keep getting cheaper while the long-term trend for the cost of both oil and gas will be upward, due to both the costs of production and market manipulation.[15]

The concern that solar power costs too much is history. Solar PV module prices dropped by roughly 75% between 2009 and 2014.[16] Today solar costs continue to plummet and have reached a point where utility-scale solar projects are outbidding natural gas generation in places like California, Colorado, and Texas on today's cost basis alone.[17] Residential PV plus energy storage systems are fast becoming cost-effective throughout California and are predicted to become commonplace throughout the country between 2020-2030.[18]

In fact, if we look at what each type of energy really costs the American people, renewable energy is by far already our lowest cost source, and with more investment in it, and ongoing improvements in technology, costs will continue to decrease. A key factor here is that the cost to produce electricity with solar panels and wind turbines does not rise each year. The cost of the fuel required to extract and convert oil and coal into useable energy accounts for most of the cost of their production, and that cost consistently rises, with occasional short-term dips.

With solar and wind power, once the infrastructure is built, the cost of generating the energy is largely fixed. As long as the sun rises and the wind keeps blowing, the fuel costs remain the same: zero. Renewable power is a comparative bargain over time because it is inflation proof. In fact, as the initial cost is paid off, the cost actually goes down each year. And the initial cost of building renewables installations has gone down dramatically in recent years. As anyone who took Economics 101 will know, those costs will keep going down as more plants are built, the market develops further, and competition heats up. Technological innovation will also keep bringing them down.[19]

We need only consider, by way of comparison, the enormous payoffs of the large dams we built a half a century ago, which today provide us with electricity that costs less than one cent per kilowatt hour to produce. The solar and wind power plants we build will produce comparably low-cost energy once their construction costs are paid off, and they will last decades past that point.[20] Just as with our hydroelectric dams, in those following years, they will provide electricity that is almost free.

If we made investments on the basis of the cost over the life of the plant, as we should, all new power plants would be renewable, even if there were no climate change. That's not even factoring in the costs of the damage of our climate, the health costs of air pollution, such as asthma and lung disease, and other externalities, which the marketplace doesn't take account of. One estimate of these costs is that they will reach 865 billion dollars per year by 2050.[21] And while much complaint has been made about governmental subsidies to the renewables sector, we must not forget that the fossil fuels industries receive subsidies that good estimates indicate are twenty-five times larger than those for renewables.[22]

There are leaders who know the value of transitioning to a renewable energy infrastructure. Cities, provinces, states, and countries from across the globe are beginning to implement 100% renewable goals for all of their electricity, heating/cooling, and transportation by 2050.[23] Denmark's energy strategy is to achieve 100% renewable electricity and heat by 2030 and 100% for all energy sources, including transportation by 2050. A bill was passed in the Hawaiian state legislature and signed into law in June 2015 by the governor David Ige to have the electricity in the state run completely on renewables by 2045.[24] As we write this book more than fifty cities have announced the goal of being 100% renewable by midcentury or earlier including, San Diego; San Francisco; Sydney, Australia; Copenhagen, Denmark; and Vancouver, Canada.[25] President Obama, also knowing the importance of getting to near zero, pledged to reduce greenhouse gas emissions 80% by 2050 compared with 2005 levels.[26]

The long-term steady increase in the cost of fossil fuel favors not only building renewables plants going forward, but over time replacing all the machines that now use fossil fuels so that they run on renewable electricity. As Saudi Arabian Sheik and former Saudi oil minister Ahmed Zaki Yamani once said, "We didn't end the Stone Age because we ran out of stones." We must not let the fact that we still have adequate supplies of fossil fuels, in particular the boom in natural gas, deter us from making the transition. As we will show, natural gas contributes as much or more to greenhouse gas buildup than the burning of coal—in total, considerably more.

The Remarkable Power of Increasing Efficiency

Converting our national ground transportation fleet, including cars, trucks, buses, and railroads to one that runs directly on renewable produced electricity is a giant step to an all-renewables future, a conversion which would earn us impressive savings from efficiency. Electric vehicles are much more efficient than combustion engines and effectively utilize about 59–62% of the electrical energy from the grid to power at the wheels, while conventional gasoline vehicles only convert about 17–21% of the energy stored in gasoline to power at the wheels. Thus substituting electricity for oil and natural gas will dramatically reduce total energy needs in America and save American consumers many billions of dollars a year.[27] This is especially true because the price of electricity is regulated on a cost basis—meaning that it must be priced at an amount based on the cost of generation determined by regulators—while the prices of oil and gas are subject to market manipulation, such as by the OPEC cartel, that can keep them artificially high.

There is no doubt that America can build an all-electric train system over the next thirty-five years. Electric trains are commonplace throughout the world and are demonstrably more efficient than diesel-powered ones. And it is now equally clear that electric cars, trucks, and buses are commercially viable and can be the vehicles of the future, a future that can begin right now with the proper mandates and incentives.

Another important source of greater efficiency is the ongoing dramatic improvement in ordinary consumer products and appliances such as, lighting, refrigeration, air conditioning, heating, and the buildings themselves. The LED light is a good example of energy efficiency. Heat pumps, a renewable heating and cooling source, can replace natural gas and oil furnaces and will also be an important part of increasing efficiency in the home.

All together an all-electric energy supply is projected be about 39% more efficient than the present system. The vast majority of this efficiency will be due to converting our system from running on fossil fuels to electricity. The remainder will be from energy efficiency. The greater efficiency should more than offset increased energy use due to growth in the economy in the decades ahead.[28]

Anyone who doubts that such an offset of increased demand is possible should take note that from 1973 to 1985, the United States GNP grew 40% with near-zero growth in energy use.[29] We did it in large part by passing a law in 1975 that required automakers to build cars with better gas mileage. And many other laws and regulations mandated that buildings be better insulated and that utilities make investments in efficiency.[30]

When Dave Freeman, a coauthor of this book, took over as general manager of the Sacramento Municipal Utility District (SMUD) in 1990, he set the goal of zero growth in electricity for the 1990s, even though the population and the economy were projected to grow at a healthy pace. It was not just a goal on paper. The necessary efficiency programs to make it happen were initiated. In one program SMUD paid people to trade in their old refrigerators for very efficient new ones. SMUD destroyed thousands of these electricity wasters and shipped the dangerous chemicals inside them to DuPont. Another program planted a million trees near homes to provide shade, which led to a substantial reduction in air conditioning usage.

It all worked. People would stop Dave on the street and give him a big hug and say, "Thanks for cutting the electric rate." SMUD didn't cut the rates at all; it cut the people's usage and their bills were lower.

Hydrogen in the Mix

Whatever happened to the promise of hydrogen power? Where does it fit in the mix of renewables?

A decade ago, the auto companies fell in love with a device called the hydrogen fuel cell. The exciting prospect was a car that ran on hydrogen, the most plentiful element on Earth and emitted only water vapor. When burned, hydrogen recombines with oxygen in the air to form water (H2O) vapor. No carbon or other greenhouse gases are released. Imagine if the only emission from tailpipes of motor vehicles was a faint mist of benign water vapor? So promising was the prospect that President George W. Bush declared his dedication to pursuing a hydrogen future in a State of the Union address.

What happened? Though one-million-dollar demonstration-model hydrogen fuel-cell cars were built that proved the technology works, there are only a handful of these cars available to the general public in 2015 and hydrogen fuel-cell cars are not going to be available in mass production by most manufacturers for years to come. A few years ago, President Bush said that, "Performance and reliability of hydrogen technology for transportation and other uses must be improved dramatically. . . ." This is still true, however, since that time dramatic improvements have been made. Hydrogen-powered bus fleets are on the road around the world and Toyota recently announced that it plans to release a hydrogen fuel-cell car in California by 2016 for roughly $45,000. Nevertheless, there is no hydrogen infrastructure and cost is a challenge.[31]

We are emphasizing an All-Electric America, but we also advocate the launch of a Manhattan Project-style research effort to find a way to separate hydrogen from water using the combination of the heat of the sun directly and a benign catalyst. The promise is too important for us to fail to make a major push. Hydrogen produced from solar power directly can be a huge new GHG free source of energy.[32] But even today we can convert solar power and wind-powered electricity to hydrogen so that it becomes a form of storage, enabling us to put sun and wind power in the gas tank and feed it into the electrical grid. Hydrogen can also replace fossil fuels for all industrial uses, power airplanes and ships, and can be used for heating and cooking. What's more, hydrogen is superabundant; in fact, it's the most plentiful element on earth and in the universe. It makes up about 75% of all matter.

The bad news is that there are no hydrogen wells; it doesn't exist in its pure form in nature. It must be created by separating the hydrogen molecules from the other elements in water or fossil fuels, and doing so today takes additional energy. Creating hydrogen from fossil fuels or with nuclear-generated electricity, with the goal of producing clean energy, is therefore like a dog chasing its tail, or putting lipstick on a pig.

Fortunately, research is underway to discover how to produce hydrogen more efficiently and a breakthrough in those efforts could help usher in the true golden age of plentiful clean energy that is upon us. Renewable hydrogen produced by separating it from water with the heat of the sun would be a viable replacement for all the forms of energy we now use. It could completely replace both fossil fuels and atomic power. Renewable hydrogen alone could end the climate crisis.[33]

Rather than pouring more money into nuclear power, subsidizing research, and building new fossil fuel plants, we should dedicate those funds to hydrogen research, while simultaneously pushing ahead with vigor on the development of solar and wind generation, energy storage, electric vehicles and other renewables, such as geothermal energy.

A View of the Future

An "All-Electric America" in 2050 would give new meaning to the electric power industry's slogan from the 1950s that we should "Live Better Electrically." The most important added value would be that the danger of runaway global warming would be averted if all nations followed our lead. But there would also be many other important benefits.

The air in American cities would be clean, and the cost of fuel—electric or hydrogen—for our cars would be about the equivalent of paying one dollar per gallon for gas. The solar power panels on just about every roof would be sending electricity to minigrids in neighborhoods. A smart meter and smart panel would turn off your lights and control your thermostat for optimal savings. Your overall energy bill would be lower, even though you'd be using more electricity, in part because you would be producing some, or all of it, and the price will have been regulated to remain steady. You wouldn't need to stop at filling stations anymore because you'd be charging your car at home while you sleep or whenever it is parked in the garage. For shorter trips where you might normally take an airplane, you would be able to ride on high-speed trains at cheaper prices and about as fast.

The country would be free from the need to support dictators in the Middle East. We'd have no need to despoil our environment further by continuing to mine for coal, drill for more oil, build enormous pipelines for its transport, and build fracking wells.

An "All-Electric America" would preserve "America the Beautiful" while also maintaining the high-energy-use way of life we enjoy and making us healthier—all at a lower cost.

The Forces of Resistance

Most utility companies are reacting defensively to the recent advances in renewable power generation. They complain that the electricity generated by large solar fields and wind farms and solar-equipped homes is unreliable and terribly expensive. Some have even called it "junk energy" because of its variable nature. To dissuade people from installing panels, some utilities pile on as many charges as possible in agreements with solar customers. Some require solar users to pay what the companies argue is their share of costs for maintenance of the grid in addition to still paying the full cost of the power they purchase from the utility, so they are effectively charging twice for maintenance. Many utilities also refuse to pay for surplus power from the solar owner.[34]

This rearguard reaction is understandable, though not defensible.

The solar threat in particular is the first seriously disruptive technology the industry has faced. Utilities will be forced either to accept a sea change in their business model or to adjust to the loss of more and more customers as they bypass the utilities. The solar bypass threat is so real in the West and the Sunbelt that some analysts are already warning about a "death spiral" and "stranded assets," meaning power plants and other expensive infrastructure that may not be needed any longer.

It does not have to be like this.

Storage Is the Solution

One of the most exciting developments in the energy industry in recent years is the improvement of energy storage technology. The total storage capacity today is comprised of a mix of mature technologies, technology in demonstration projects with strong developmental potential, and brand-new, cutting-edge technologies.[35] Fortunately, even as many utilities dig in their heels, others see the value of these technologies and are investing in them, especially in California, where the public utility commission is requiring 1,300 MW of storage capacity to be purchased by utilities to "get started." They understand that storage is the essential partner to wind and solar that will make renewable power reliable and efficient and allow them to widen coverage.[36]

Both the energy advisory firm Navigant and the investment firm Barclays are reporting that storage prices are plummeting. Deutsche Bank expects the cost of storage to decrease from fourteen cents per kWh today to approximately two cents per kWh within the next five years, with lithium-ion batteries achieving 20-30% yearly cost reductions. Goldman Sachs is betting good money on prices coming down, investing $40 billion in the renewable industry.[37] They are not alone. Other investment companies, as well as think-tank centers such as the Rocky Mountain Institute, also predict lower costs and see the combination of renewable generation and storage for residential customers as the most likely scenario for the future.[38] The announcement of lower prices for batteries by Tesla on April 30, 2015, is resulting in making these predictions come true much sooner than expected.[39]

An old Chinese adage tells us that every threat is also an opportunity. Electric utilities have a choice: They can continue fighting a losing battle and frustrating customers who want to install solar panels. Or they can get aboard and offer to install rooftop solar as efficiency measures. Better still, they can start promoting heat pumps and electric cars that can utilize the electricity they purchase from large solar projects. The electric industry once advocated for a massive expansion of the electricity infrastructure. It was the industry's goal for a couple of decades starting sixty-five years ago when they expanded into rural America and began promoting the "All-Electric Home" in which one could "Live Better Electrically." That forward-looking vision must be restored.

Natural Gas—The Route to Climate Hell

Only a few short years ago, most environmentalists and energy enthusiasts alike were excited by the prospects of natural gas. The promise of a fuel that produces half the carbon dioxide of coal when burned appeared to be the answer to our climate problems.

The public debate about the dangers of natural gas has focused primarily on the possible damage caused by fracking—hydraulic fracturing—in which the gas is released from deep within rock formations by injecting chemically treated water at extremely high pressure into cracks, further opening them and releasing the gas. What is not widely understood is that the gas released is made up largely of methane—the most potent of the greenhouse gases—and that not all of the methane being pumped out of the earth is burned; a significant amount leaks into the atmosphere from the wellheads as well as from the old pipes that transport natural gas to almost seventy million homes and two hundred thousand industrial plants.[40] The fundamental problem is that the federal government counts carbon, but not methane, when Mother Nature actually does. We really don't know how much is emitted.

The latest scientific studies have shown leakage rates of methane from 1.5% of the total natural gas produced to as high as 10-17%.[41] Leaked methane is 120 times as damaging to the climate as carbon dioxide when released. Its impact reduces with time, but is still 84 times as damaging over twenty years and 34 times over one hundred years.[42] So when you add the methane to the significant carbon emissions from natural gas, it becomes clear that natural gas is as bad, or worse, than coal as a greenhouse gas emitter.[43]

In order to address the problem, President Obama created a task force to draft regulations to control and measure the leakage, but, to date, the recommendations made are purely voluntary and even when finally implemented will apply only to new construction after August, 23, 2011. They also do not cover massive leakage in the old natural gas pipes beneath our cities. For the foreseeable future any shift from coal to natural gas will not reduce total emissions.

However, even if all of the leaks were sealed, natural gas would not solve our climate change problems. Because we now use twice as much total natural gas as coal, we actually produce almost as much carbon dioxide alone from our total natural gas use as we do from coal.[44]

Substituting all of our coal plants for natural gas plants in the electricity industry would still result in our producing enough carbon dioxide to put us squarely on a path to disaster. Without leakage we are on a road to ruin and with leakage it is a whole lot worse.[45]

The bottom line is that the continued use and extrapolation of natural gas is the road to climate hell.

"All of the Above" Is a Road to Ruin

Despite all of the public campaigning about the urgency of the global warming problem, the prospects of containing warming, and eventually stopping it, are dimmer every year that we fail to vigorously pursue an all-renewables strategy.

The best way to think about the climate challenge is to consider our emissions as another form of national debt. Because the carbon and methane that is causing climate change stays in the atmosphere for many years, all the greenhouses gases we emit each year add to that buildup. Just as with our staggering national financial debt, every year the buildup grows larger and the task of halting warming becomes more daunting. We must achieve near zero greenhouse gas emissions by the year 2050—this gives us about thirty-five years. We should think of zero emissions as achieving a "greenhouse balanced budget."[46]

During the 2009 Copenhagen Conference, the United States and other participating nations agreed that we should not heat our earth more than two degrees Celsius above preindustrial levels.[47] They have chosen this temperature as a target because a wide range of high-quality scientific studies have indicated that if we are to contain warming, stopping it from escalating out of control and causing massive planet-wide destruction, we must keep the temperature rise within two degrees Celsius. In order to achieve this, the entire world must emit no more than 270 gigatons of carbon from greenhouse gasses into the air this century. Two hundred seventy gigatons is our remaining world budget and if we stay on our current path we will blow past our budget by 2033![48] [Note: "The 270 gigatons of carbon accounts for the impact from carbon dioxide as well as other warming agents such as methane. It is derived from carbon dioxide equivalents (CO_2eq) — e.g. the concentration of CO_2 that would cause the same level of warming as a given type and concentration of different greenhouse gases.][49]

The only way to achieve a balanced budget is to reach zero greenhouse gas emissions by the year 2050.[50] This means that we have to keep our fossil fuels "in the ground."[51] Fossil fuel companies surely want to burn it all.

The tragedy of the existing climate debate is that not only our political leaders, but most of the environmental community as well, are not even proposing measures that would ever achieve a balanced greenhouse gas budget. As a result, we are getting deeper "in debt" every year.

The "All of the Above" policy of the Obama administration, which favors expanding the use of natural gas, oil, and nuclear power, is fundamentally flawed. That approach will not only fail to achieve zero emissions, it will seriously impede any real progress toward that goal. The core problem with this approach is that it is geared toward achieving carbon reductions by shifting from coal to natural gas rather than by realizing that all new energy production must be of renewables. "All of the Above" instead ignores methane releases and funds improvements in nuclear power and "clean coal."

The endorsement of the pursuit of "safe nuclear power" is a fool's errand. Nuclear power plants, which are piling up radioactive waste with no safe disposal plan in sight, are a failed experiment that, as we will describe in greater detail, will not only remain highly dangerous but much too expensive. Nuclear power plants are also a path to making nuclear bombs, an example we must stop setting. The Obama administration has argued for a new subsidy for research into improving nuclear technology, but that money would be much better spent on promoting accelerated adoption of renewable generated electricity.

Another cornerstone of the "All of the Above" policy, to move to "clean coal" by funding the installment of equipment on old plants that captures carbon emissions and stores them in the ground, is unfortunately another false god. The costs are so high and carbon storage is so uncertain that no such new coal plant is even planned that is not part of a plan to produce more oil. Unfortunately, the Obama administration has used the patently false label of "clean coal" and supported that falsehood with large sums of federal dollars.[52]

The Leadership Gap

Rapid progress toward an all-renewables future is being stymied not by lack of technology, or even by cost or market demand, but by lack of vision on the part of our political and business leaders, and lobbying and persuasive advertising by the oil, gas, coal, and nuclear industries. President Obama, environmentally minded political leaders, and most of the major environmental organizations have been promoting both the "green revolution" and the "brown surge," supporting both renewables and the continued use of fossil fuels. They have failed to hammer home the message that a completely renewable future will be lower in cost, as well as necessity if we are to halt global warming, much less propose programs to make it happen. This is despite the fact that a long-sought bipartisan goal of U.S. energy policy has been to achieve energy independence. An all-renewable supply is the best way to do so.

Many political leaders take pride in saying they believe the climate science and are concerned about climate change, unlike the folks who deny that the problem exists. But even those concerned about climate change are not proposing actions that will control the greenhouse gases to reach our ultimate climate goals before it is too late. We respectfully suggest that the failure of most of our political and environmental leaders to propose actions that will, in fact, reduce total greenhouse gas emissions sufficient to stay under the 2°C limit puts them in a group that can be considered "intelligent deniers." Let us be specific.

The greenhouse gases from our energy infrastructure are emitted by four major sources:

- Electric power production
- Heating of buildings
- Industrial uses
- Transportation

However, major policy actions to assure that emissions are reduced to meet climate goals are only being discussed for electric power. Heating of buildings and transportation, fueled by oil and natural gas are not even part of most of the debate. To be sure, mandates are making new cars more efficient and a few electric cars are now being sold. Yet, both the "intelligent" and the "regular" deniers in effect are saying, "frack baby frack," bragging about discovering more oil and gas, while the scientists tell us we can't safely burn more than 25% of what's already been discovered.

The science requires that over the next thirty-five years—starting no later than NOW—we reduce to zero our use of oil and natural gas. At the same time climate leaders ignore the science and support the largest cause of our climate problem, the burning of petroleum and natural gas.

Let's not be deceived. The majority of greenhouse gas emissions are from natural gas and oil. They are well over 50% of the problem. And climate action leadership has yet to show the political courage to start reducing, not enlarging, the use of oil and natural gas that is already reaching havoc on earth.

The great irony of this situation is that cost-effective technology exists to replace most all of the oil and gas we use today for all of our transportation, industrial, and heating needs.

Investment in the Future

As we will show in more depth in the following chapters, the inescapable and stark truth is that all approaches that detract from investment in renewables are impediments to the much more realistic, efficient and cost-effective goal of an all-renewable, greenhouse-gas-free supply. There is no question that the costs of investment in the infrastructure and the research required are significant, in the trillions of dollars. But they are well within what we are capable of and the costs of failing to make the investment dwarf them. Making the transition will create millions of jobs and provide us with a reliable, entirely independent energy supply that is cheaper in the long run.

We must appreciate that the choice we face today is both a challenge and an opportunity. And it is the American public who will demand we seize that opportunity. Later in this book we will propose a program for outlawing the building of any new fossil fuel plants in the U.S., electrifying the railroads, and implementing an all-electric energy supply by 2050. At the heart of the plan is the eminently practical goal of steadily reducing the use of fossil fuels—by 3% a year—while building alternative greenhouse-gas-free electricity generating capacity at a pace to meet our future 2050 energy needs.

We know full well that this program would not be adopted by the U.S. Congress, or proposed by the president, this year, next year, or any year if left to their own devices. But if the test for responding to climate change is what the U.S. Congress will pass, or the regulations the presidents will put in place of their own volition, we are surely doomed to failure. Yet individuals, cities, and states are acting now. The driving force is concerned citizens who demand plug-in electric and hydrogen fuel-cell vehicles; who purchase only ENERGY STAR appliances; who install electric heat pumps and solar water heaters in their homes; who demand green power from their utility companies; who install solar panels on their roofs and storage capacity as backup; and who lobby and elect senators and representatives to enact requirements and incentives to assure attainment of the all-renewable All-Electric America.

That is why we've written this book; to inform the engaged public about how hopeful the prospects are, how substantial the payoffs would be, both for individual households and the public at large—not to mention the whole planet—and how urgent the need to change course is. The best estimates indicate that achieving zero emissions will take thirty-five years from when we begin the transition in earnest, and thirty-five years is all the time we have.

In the past, when this nation faced up to a deadly threat or an exciting challenge, we took action collectively. If it was poisonous, we outlawed it, as with DDT. If it saved lives, we mandated it, as with seat belts and airbags. And if it was a challenge, like going to the moon, we funded NASA to build a spaceship to get there. Greenhouse gases are our greatest challenge yet. We will only meet the challenge if the public demands vigorous action by our government now.

CHAPTER 2

Solar and Wind
Can Electrify Everything

Mother Nature delivers a superabundance of energy to the earth, free of charge. Every day, the sun delivers five thousand times more energy to the planet's surface than the whole world consumes. According to a report by the DOE "the amount of solar energy falling on the United States in one hour of noontime summer sun is about equal to the annual U.S. electricity demand." Many people believe that solar power is only feasible in the warmer climates, but as the report highlights, "every region of the contiguous United States has a good solar resource."[53]

Each continent on the planet also receives enough energy directly from the sun that, were sufficient collectors installed, the solar energy captured would negate the need for any other source of fuel. We would no longer need to burn fossil fuels, mine coal, frack for gas, or construct dangerous nuclear power plants—ever again.[54]

Yet in 2014, solar power accounted for only 0.4 % of the total energy used in the U.S and only 0.5 % of utility-scale electricity generation through March 2015.[55]

The good news is that a true solar revolution is well under way that is reducing the cost of solar power below the cost of electricity from coal, natural gas, oil, and nuclear power, when calculated on the basis of the costs over the life of the plant.[56] The installation of PV panels is likely to continue apace as their prices have plummeted in recent years. Heavyweight investment bank Goldman Sachs is expecting them to continue to decline by 3% per year and Deutsche Bank predicts prices will fall 40% by the end of 2017 with the greatest cost reductions in the residential sector.[57]

The solar revolution is real. Companies like Solar City will install solar panels on your roof and sell your solar power at a price lower than current rates from your utility in California, Hawaii and several other states. PV solar panels in 2015 are selling in the market for as low as fifty cents a watt, representing a 60% fall in just three years. With that market price, large solar farms can offer power for under six cents a kilowatt hour, which is very competitive with new fossil fuel or nuclear plants.[58]

In fact, surprisingly even lower utility scale PV prices are appearing on the market. Jim Hughes, CEO of First Solar, has said that, they are "... regularly bidding in at ... five- and six-cent power" and "... beginning to see four- to five-cent power." He expects fully installed solar to sell for one dollar a watt by 2017 in the western United States.[59] Austin Energy in 2015 procured 1.2 gigawatts of solar for less than four cents and expects prices in the future to fall below twenty dollars a megawatt-hour (a price lower than coal), so Hughes's projections may become reality even sooner.[60]

Over a year's time, consumers who install solar panels in an average home in many states can save two hundred to five hundred dollars. And those savings increase every time the utility raises its rates, which happens fairly often. As for the states that have poor sun, many have strong wind power, which has the same long-term cost advantage.

Solar and wind power, backed by energy storage, will stabilize the cost of electricity in the years ahead for existing uses. That will be a huge savings compared to the steady increase in price that is the current utility pattern. Yet the greatest cost saving to consumers will be in substituting electricity for oil in transportation and for natural gas, propane, and heating oil for heating homes, office buildings, and factories. Here we can substitute the equivalent of one-dollar-a-gallon electricity for two- or three-dollar-a-gallon gasoline. And a heat pump will undercut natural gas and oil heat today and be dramatically cheaper in the years to come.

In announcing a deal struck by Xcel Energy to provide the state of Colorado with solar power, David Eves, CEO of Public Service Co. of Colorado, remarked that PV-generated electricity is now cost-competitive with natural gas-fired generation in that state and had "made the cut . . . purely on a price basis . . . without considering carbon costs or the need to comply with a renewable energy standard . . ."[61]

The installation of batteries and other devices to store solar energy for use at night is also now a commercial fact. In November 2014 the Southern California Edison Company announced the purchase of 250 MW of storage capacity. The state utility commission has ordered all California utilities to purchase 1,300 MW as a beginning. In Spring 2015, Elon Musk unveiled its Tesla Powerwall residential battery and a utility-scale battery, the Tesla Powerpack.

The lower cost of residential solar capability combined with better technology for storage is driving rapid adoption by households, beginning with the Sunbelt, and spreading rapidly. Increasing numbers are poised to leave the grid, provoking some utility executives and regulators to warn of an electrical "death spiral." The threat, or more appropriately, the challenge, to utilities from companies offering solar plus battery storage to their customers is real.[62]

Our rooftops alone are estimated by the Department of Energy to have a capacity of about 664 GW of generating capacity, a little over half the total U.S. electric capacity.[63] Barclays bank analysts have predicted disruption of the utilities business and downgraded the bonds of the entire electric sector of the U.S. from high-grade to underweight.[64] The best way for the utilities to meet the challenge is to join the revolution, and the best way for the American public to convince them to do so is to install solar capacity in their houses.

Solar power in the home has a long history. The adobe homes of Native Americans in the Southwest are an example. For all new homes being constructed, a range of passive solar features can be incorporated, such as south-facing concrete walls that absorb heat and trees that cool a home naturally.

Anyone building a new home could ask their architect about the possibilities and at the least should insist on installing rooftop solar. For those already in their homes, solar panels can easily be installed on your existing structure. And roofs are not the only place solar panels can be installed. You could also look into solar-enabled shingles for your roof. And many communities will establish neighborhood solar plants with microgrids.

Clever designers and architects are continually finding better ways to incorporate solar into designs, such as on awnings. If you haven't explored the options, you will probably be pleasantly surprised by how many companies in your area that can do the job will come up in a Google search for "solar products and services."

In the short term, the solar power you harness with battery storage will protect you from outages and cost fluctuation. In the long term, it will save households a great deal of money— several hundred dollars a year for many, many years.

For those who have already equipped their homes, if they're not being paid the retail price for their surplus solar power, they should call their state public utility commission and complain and should write to their state regulator insisting that selling to their utility at retail price be made mandatory. Swaying the companies and the regulators will take time, but the more customers who make the move, the more leverage they will have.[65]

Big Solar—Land Aplenty

Large-scale solar generation is well under way in California and the Southwest. The capacity to produce thousands of megawatts has been built and capacity for thousands more is under construction or in the serious planning stage.

The argument against large-scale solar comes from people who don't want any development in their neighborhood or line of sight. They fear solar-field eyesores all over our communities and countrysides. Nothing could be further from the truth.

The environmental impact of solar can be minimized by placing panels on a wide range of structures in our already built environment. We have vast acreages available on rooftops, both on homes and on commercial buildings. Also available are parking lots, parking structures, and disturbed land once dedicated to industrial or other commercial uses, such as shopping centers and factories that have closed. The decentralized solar in or near already built-up communities is a favorite option.

But if we are to become all-renewable in thirty-five years, we need to also build where the sun is hot and the days are long—our deserts. We need not disturb pristine land; more than enough is already disturbed or environmentally suitable and available for solar development.

If, for example, we install solar PV on just 0.16% of our land we will generate the equivalent of 100% of our total 2013 electricity in the U.S. with PV alone. And 0.62% of total land cover would be needed to generate the equivalent of our entire projected 2050 energy capacity needs for a completely renewable energy infrastructure.[66] The land that will be needed for an actual all-renewable energy infrastructure, however, is even less than this because resources that use less land such as wind and rooftop solar PV will be part of the energy mix.

Solar installations don't scar or contaminate the land they're on, as coal mining, oil drilling, and nuclear power plants do. By comparison, extraction and transport of oil and coal and fracking for natural gas are ecologically devastating. Oil spills are commonplace and kill large numbers of marine birds and animals in our seas every year. Coal mining kills miners and the land, not once, but every year. Fracking is contaminating our water tables with its cocktails of toxic chemicals. The land used for nuclear energy plants becomes contaminated with toxic chemicals. And the risk of accidents, such as that at Fukushima in Japan recently, is always present with the threat of large-scale and permanent destruction of large swathes of territory and exposure of the local population to radiation. The partial meltdown at the Three Mile Island plant in 1979 could have left an area larger than New England permanently contaminated with radioactivity, just as the 1986 Chernobyl accident did in the Ukraine and Fukushima did in Japan. And don't forget the land disturbed to mine uranium and the health risks for those miners.

In sum, our land will be much better off with solar panels installed on it than by continuing to ravage it by mining and drilling for fossil fuels.

The Possibilities of Wind Power

As we embark on the road to reducing greenhouse gas emission and oil imports, wind power will have an essential role right away. Fortunately, the electric utilities have been receptive to wind generation. Al Gore, in a Rolling Stone article, explains how Texas has become our largest wind producer, how it has become cheaper than new coal, and that "nearly one-third of all new electricity-generating capacity in the past five years has come from wind. Installed wind capacity in the U.S. has increased more than fivefold since 2006."[67] More than 65,879 MW of wind-generated electricity had been transmitted in the country as of December 2014.[68] And we've only scratched the surface of what we can do. In 2014, wind power produced roughly 4% of U.S. generated electricity in the country, when it is capable of producing 100%.[69]

The Department of Energy has estimated that the country has a wind capacity of 11 TW from onshore wind farms and 4.2 TW from offshore installations. This is 8 times the total U.S. electric power capacity today and roughly 2.5 the total capacity required for all of our energy needs in 2050. Just as our deserts are gold mines for solar power, the central region of the United States is a veritable Saudi Arabia of wind, one of the best energy resource sites in the world.[70]

Much of our current capacity has been built with the support of government subsidies and tax incentives, but wind is becoming competitive even without credits or subsidies. Utilities are willingly contracting for wind power to diversify their portfolios, and not only in response to mandates and public opinion. They have realized that wind power is a vital link in the chain to ensure they have low-cost energy to offer in the future.[71]

In fact wind power often takes the place of nuclear-, coal-, and natural-gas-generated electricity in the late evening or at night when demand is low in the wholesale markets. This actually suppresses prices and can benefit the customer because wind farms are dispatched ahead of fossil fuels since their marginal costs are zero — e.g. renewable fuel is free. This, however, can make it difficult for existing fossil fuel plants to compete and make money on the wholesale market. This problem will disappear as we approach a 100% renewable and flexible power option that includes storage.[72]

The U.S. Congress and states often threaten to phase out the existing tax credits that have really helped wind power get started. This uncertainty is yet another reason we need requirements instead of the volatility of the market to reach a zero GHG goal.[73]

Examples of regions and states seeing wind compete on an even playing field include, but are not limited to Texas, Colorado, Oklahoma, and Midwestern states. Additionally, utilities in Georgia and Alabama that are buying low-cost wind from Oklahoma, Kansas, and New England announced in 2013 that wind is cheaper than gas-fired alternatives.[74]

In December of 2013, wind projects in the Midwest, which enjoy not only high wind speeds, but low construction and labor costs, won bids over natural gas and coal producers in competition for new contracts. The Midwest has seen agreements with wind farms for wind electricity priced as low as twenty-five dollars per MWh, compared to the cost of electricity from an existing gas plant of thirty dollars per MWh and about sixty dollars per MWh for a new gas plant. The comparison to coal-generated electricity, which runs between twenty dollars per MWh and twenty-five dollars per MWh,[75] is also impressive.

In summary, with tax credits, wind power is a serious competitor to fossil fuels even on the basis of short-term costs. There are many places wind can compete even without tax credits. However, this is not enough and the continual threat of the expiration of the tax credit creates uncertainty and inhibits investment (As of June 2015, Congress had not yet voted to extend wind's tax credit into 2015).[76] The looming expiration in 2016 for solar and the threatened expiration for wind created a cloud over their economics in the short term. This is true even though it is very clear that over the life of the renewable power plant it is the lowest-cost option. For that reason we do need the force of law to do the right thing.[77]

In addition, wind projects face citizen opposition. One key group arguing against them is animal activists, whose primary concern is that birds get caught in the turbines. But the truth is that wind turbines impact animal life, as well as land, very lightly compared to other human intrusion on the natural environment. And intrusion is quite minor by comparison to fossil fuels and generation of nuclear power. The American Bird Conservancy found that the number of bird deaths per year from wind turbines is 573,000 per year. This is far less than deaths caused by flying into windows, estimated at 300 million to 1 billion per year, and getting tangled in transmission lines, at 175 million per year.[78] Other estimates calculate that wind turbines cause .27 avian fatalities per GWh of electricity generated, while nuclear power plants cause 0.6 fatalities per GWh, and fossil-fueled power stations (coal, natural gas, and oil generators) are responsible for about 9.4 fatalities per GWh.[79] So we will actually save millions of birds if we switch from fossil fuel and nuclear power to wind.

Concern has also been raised about turbines using up valuable land for grazing and agriculture. But the distance required between turbines permits both planting crops and grazing herds on wind farms on almost all of the land. In fact, only a tiny fraction (~0.0025%) of the land in a wind farm is actually occupied by the turbines themselves.[80]

Much concern has also been voiced about the impact of offshore wind power on marine life. Offshore wind has been developed extensively in Europe and studies are finding little long-term impacts on wildlife. A study conducted in Denmark, for example, found minimal impact and concluded that, ". . . under the right conditions, even big wind farms pose low risks to birds, mammals, and fish . . ." The key phrase is "right conditions." Wind farms need to be designed and built with the help of avian and marine experts and other specialists. And there are routes that need to be left alone.

The National Wildlife Federation warns that, ". . . as many as 30% of species worldwide will face extinction this century if warming trends continue. To protect wildlife from the dangers of a warming world, we must take appropriate, responsible action to replace as much of our dirty fossil fuel use with clean renewable energy sources. And wind is a key part of that task." But the good news is that there are plenty of sites where impact will be minimal.[81]

We don't have to disturb our pristine deserts or sensitive habitats. We will need energy policy planners and local citizens to join together to identify land on which turbines will have the least impact on our communities and sensitive ecosystems and animal life.

Fortunately, there is plenty of such land, no one technology will have to do it alone, and with good planning it will be possible to find the right balance. Mark Z. Jacobson, professor at Stanford University, together with his partners, find that to power all our energy needs, with wind and solar being the heavyweights, the amount of land needed is minimal and will be roughly 0.42% of U.S. land. If we include spacing of wind turbines, the area required is still only 1.6% of U.S. land.[82]

Jacobson created a model for an optimal balance of solar, wind, thermal storage and other renewable resources that can power all of the energy needs for the nation and each state individually in the year 2050. The analysis finds that roughly 31% onshore wind, 19% offshore wind, 31% utility-scale PV, 7 % rooftop PV, 7% concentrated solar power (CSP) with storage, and 5% other renewables is the ideal distribution for the country. The team also mapped out the costs, resources, and lawmaking required, as well as the jobs, energy savings, cost savings, and health benefits that will result from this transformation. They find that we will use 39% less energy, save lives, increase jobs, and achieve cheaper energy. A summary for each state can be found on The Solutions Project's website.[83]

In order to power all of our energy needs with solar and wind, an ideal system will have a large amount of storage capacity. The good news on that front is that—as we'll dive into in the next chapter—storage technology is available now, it is improving exponentially, and costs are plummeting.

CHAPTER 3

The Energy Storage Solution

One of the obvious obstacles to an all-renewable electricity system is how to assure reliable power. Large coal, natural gas, and nuclear plants can generate power day or night. Solar and wind plants cannot. For that basic reason, storing some of the solar and wind power for use when needed is the vital partner to these types of power generation in creating the all-renewable future. Today's electric systems contain a 15-20% reserve margin so that if a large plant breaks down, the lights still stay on. Utilities in regional grids have a good record of reliable power, except during hurricanes or blizzards when the power lines are blown down.

The foundation of this system is the inflexible, expensive, polluting, and climate-changing infrastructure of large coal-burning, natural-gas, and nuclear plants. That supply is complemented with smaller plants used primarily when the demand is greatest, with natural-gas-fueled peaking plants being the main source for this function.

Storage has always been an unnoticed part of our energy supply. Batteries in cars are an example we never think about. Power plants include a ninety day stored supply of fuel for emergencies. The lake of water behind a hydropowered dam is stored energy to be used when needed. And in reality the 15-20% of capacity, generation that is needed only when some of the power plants unexpectedly break down, is a very real form of storage also. Even solar thermal plants store some of their solar heat to generate electricity at night and thus function as reliably as a gas-fired plant.[84]

What is new is not the idea of storage but the need for a very different kind and magnitude of it to accommodate a power supply that is inherently part time and needs backup reserves not just for emergencies but all the time.

Today's electric system is increasingly vulnerable to disruption from the superstorms being brought on more frequently by climate change. But the fundamental problem is that today's power supply system is lacking in the storage capacity to facilitate the move to an all-renewable electricity supply.

Why We Need Energy Storage

Good estimates indicate that when more than half of the supply is coming from renewable sources, without sufficient storage, the fluctuations could destabilize the system, making it exceedingly difficult to match supply and demand.[85] The solution is to build both extra generating capacity and enough storage capacity. Those that question whether an all-renewable system can provide reliable electricity are not aware that a vast array of storage technologies exist and are already being used by utilities and consumers. Storage systems are often extremely flexible, can be used as support throughout the grid, and one type of storage unit may even be able to provide more than one service. And their costs, like solar and wind, are getting lower and lower.[86]

There are essentially three different places in the system where storage is needed:

1. *Behind the meter.* An example is a battery where solar energy from

rooftop panels is stored at a home for nighttime use. The batteries in your electric car can also store the electricity.

2. *Within the utility distribution system.* The units will be used primarily for balancing and maintaining power quality. For example, to balance the electricity flow and firm up (stabilize) the solar and wind power from the deserts and mountains, a utility could install large storage units, such as batteries, at points of generation or at each of its substations or elsewhere on its distribution system. Storage units can also be part of a microgrid to balance the distributed generation in neighborhoods or regions.

3. *Bulk storage, often near generation on the grid.* These systems can store large amounts of energy to provide energy for many hours, days, and longer and for use as arbitrage, peaking, spinning, or long-term reserves. Examples currently in use are large pump storage plants that pump water up a hill when the sun is shining and released to generate power when the sun is down. Technology such as thermal and chemical storage and compressed air are also available.

One need only to search the Internet for electricity storage companies to find commercially available technologies such as:

- Solid-state batteries, such as lithium batteries used in cars and for homes and utility-scale storage.
- Flow batteries, which store the power in an electrolyte solution for longer life and large capacity, for services such as peak storage when demand is highest.
- Flywheels, which are mechanical devices that store rotational energy and deliver instant electricity.
- Compressed air, which provides excellent long-term bulk storage reserve.
- Thermal, which captures heat and cold to provide electricity as needed for longer term and peak times.

Building more storage capacity even today is a much better way to go than building costly new fossil-fueled plants to meet demand at peaks. Storage is typically easier to site, faster to install, and produces virtually no emissions.

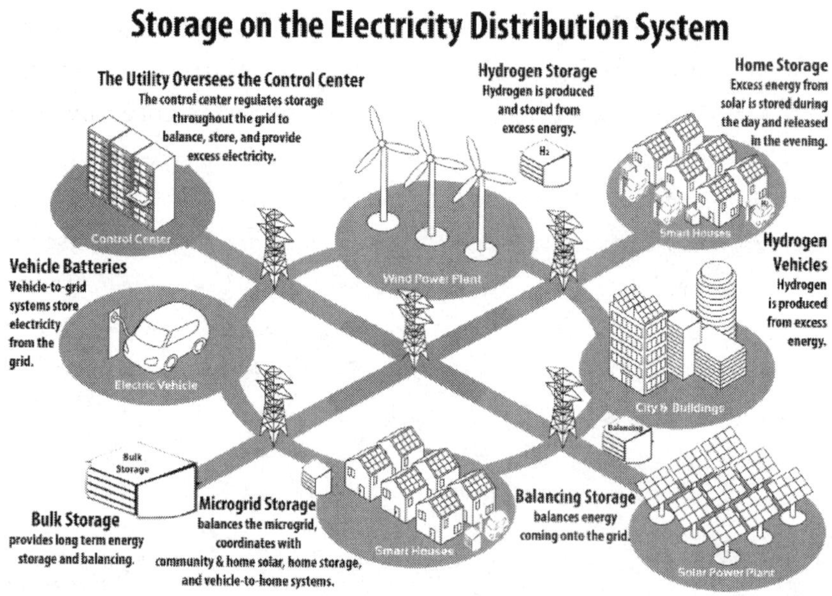

Storage on the Electricity Distribution System

The Utility Oversees the Control Center
The control center regulates storage throughout the grid to balance, store, and provide excess electricity.

Control Center

Hydrogen Storage
Hydrogen is produced and stored from excess energy.

Home Storage
Excess energy from solar is stored during the day and released in the evening.

Smart Houses

Vehicle Batteries
Vehicle-to-grid systems store electricity from the grid.

Electric Vehicle

Wind Power Plant

Hydrogen Vehicles
Hydrogen is produced from excess energy.

City & Buildings

Bulk Storage

Bulk Storage
provides long term energy storage and balancing.

Microgrid Storage
balances the microgrid, coordinates with community & home solar, home storage, and vehicle-to-home systems.

Smart Houses

Balancing

Balancing Storage
balances energy coming onto the grid.

Solar Power Plant

FIGURE 1: Energy Storage is available to meet a variety of needs and one type of storage unit may be able to provide more than one service. This schematic shows a few storage options. Only a few services and types of storage are actually shown here, other storage options such as thermal storage provided by CSP plants is not shown here. Copyright: All-Electric America.

In the short term, storage will add flexibility and will allow utilities to respond quickly to the increasingly fluctuating load from renewables supply. In the long term, when variable renewable supplies dominate, we'll rely on technologies that will store electricity and make storage for weeks or even months viable, further improving reliability.[87] Figure 1 shows an example of a few places we will find storage on the distribution system and examples of some of their functions.

A major utility argument is that natural gas generation is needed to respond quickly to fluctuations in availability of solar and wind power. What they fail to reveal is that storage devices actually do the job better.[88] They are able to ramp up and down quicker than gas turbines and are more available when there are large amounts of variable energy on the grid. Paradoxically, natural gas plants are often the first to be turned off when there is a high amount of variable energy on the grid.[89]

There are many benefits that result in storage being less expensive than one might think from the actual upfront cost. Some benefits include, customer bill offsets, avoided distribution outages, avoided transmission line upgrades, and deferred investments in excess generation capacity for backup for peak-time power. And with the continual downward trend in storage technology, such as the announcement of the new Tesla battery plant and its home and utility-scale batteries, analysts are projecting storage at two cents per kilowatt-hour or less. At this price they will be cheaper than natural gas power even without considering all of the other values.[90]

Current Storage Initiatives

Fortunately, many public officials and utilities appreciate that the development of storage is essential. A host of government-sponsored initiatives and mandates are being implemented. In particular, states and regions that are beginning to see the highest penetrations of renewable energy into the grid are determining that they must act fast to create more storage.[91]

California, New York, Oregon, Puerto Rico, and Hawaii have all made important moves. In 2014, the California government mandated that the state's energy portfolio be comprised of 33% renewables by 2020, and to support this goal, California Public Utilities Commission, lead by Commissioner Carla Peterman, directed the state utilities to reach an energy storage goal of 1,325 MW megawatts by 2020.[92] This is just a beginning in a state with some 79,000 MW of capacity, but it is good beginning.[93] One utility, Southern California Edison, awarded contracts for 250 MW of storage when they were required to contract for only 50 MW. The bottom line is that storage is now a commercial reality in California.[94] The Puerto Rico Electric Power Authority, the island's main utility, initiated a mandatory requirement to use energy storage when incorporating new renewable projects and Oregon's governor, Kate Brown, signed a mandate for five megawatt-hours of energy storage by 2020. These states are recognizing that new solar and wind must have storage as a partner.

Many utilities are coming to the realization that the best move is for them to embrace the development of storage technology. Duke Energy, PJM, and Hawaiian and Puerto Rican utilities are involved in a number of innovative projects that are combining solar and wind with battery storage to manage variable energy.[95] A large grid-scale project for this purpose currently in use involves a 64 MW battery at a 98 MW West Virginian Wind farm.[96] Hawaii Electric Company is adding 60-200 MW of storage to manage the wind and solar energy on its system. And considering the recent passage of Hawaii's House Bill (HB) 623, the nation's first bill requiring 100% of a state's electricity generated by renewable energy by 2045, this is just the beginning.[97]

The New York State Public Service Commission launched an initiative—"Reforming the Energy Vision"—that promotes energy storage and the development of microgrids, which are small-scale neighborhood grids that deliver power from renewable sources backed by storage. Long Island Power Authority, which is the publically owned utility that serves Long Island, has requested up to 150 MW of storage. The New York City utility Con Edison is adding 100 MW of energy storage.[98]

The essential truth is that as long as sun and wind are just 20% of a utility system, storage is provided by the reserves they already have for system reliability. And in the states where the renewable revolution is passing the 20% line, storage is belatedly being recognized as an essential partner.[99]

Storage technology is proving again and again to be cost-effective.[100] The hidden truth is that the cost of storage is going down with volume production and advanced battery technology. It will go down even further as the giant battery factory that Tesla is building opens, technologies continue to commercialize and are implemented, and the Chinese turn up the volume.

The good news is that storage is cheaper and a faster backup than natural-gas peaker plants even today, and certainly over the life of the plant. And that's without counting that natural gas brings on climate hell and batteries don't pollute if they are recycled.[101]

Raising the Profile for Electricity Storage

Strong mandated policy intervention will be needed in order to make storage a necessary partner to solar and wind projects from now on. Utility-scale electricity storage development has not proceeded nearly as quickly as the development of solar and wind power. It is estimated that 5,242 MW of new solar and 4,050 MW of new wind was added in 2014, while only 436 MW of storage capacity was added in all of North America during roughly the same time period.[102]

Thus far solar and wind have relied on the surplus and reserve capacity of the existing electric power grid. But as previously stated, that doesn't work when renewables supply over half of our power. The general public and government policy makers all need to know that solar, wind, and storage must be built together. We can't expect regions where solar and wind are not advancing to install storage. But in mandating that the renewable revolution must advance—a year at a time—it is crucial that new solar and wind projects include the storage to make that renewable power reliable.

What is needed is to enlarge the mandates that require a growing percent of renewable electricity to require sufficient storage (and not gas peakers) to assure the reliability of our future electric power.

In order for that to happen storage needs to be advocated in the same breath as solar and wind. Thus far most public officials and environmental organizations who claim to be concerned about climate have failed to recognize the vital role of storage as the vital partner of the climate solution.

Henry Kissinger once said that issues of war and peace are too important to be left to the State Department. Perhaps the same can be said about climate change—it is too important to be left to political leaders or many environmental organizations who tend to follow these days, not lead.

The serious efforts by utilities of their own accord are very hopeful signs that the tide will turn. The pressing question is, will it do so fast enough? This is where customer demand is again so important.

Let the word about the renewable threesome—sun, wind, and storage—become widespread public knowledge. Perhaps this book will inspire the media to explain to the public that with storage—that's available and affordable—the sun can shine and the wind can blow all the time!

What also makes the prospects of an all-renewables-plus-storage system brighter is that we can make such great strides not only by storing more energy, but by using less of it. In the next chapter, the substantial advances we can make in energy efficiency are explored.

CHAPTER 4

Efficiency: The Low-Cost Partner to Renewables

I t is gratifying to observe that companies and policies promoting energy efficiency have gained across-the-board support from political parties that differ on almost every other issue. Government officials and representatives on both sides of the aisle, many utilities and other energy producers, the business community, and consumers all agree that conservation is good. Indeed, on April 23, 2015, Congress put aside its partisan bickering long enough to pass the Energy Efficiency Improvement Act, legislation aimed to encourage the increased use of high-efficiency technology.[103]

Getting to this broad consensus that energy savings is a good deal has taken a long time. In the early 1960s, when Dave served as the executive assistant to the chairman of the Federal Power Commission, the idea of saving energy was actually widely considered anti-American. He remembers that when the staff presented projections of U.S. electric demand for 1980 to his boss Joseph Swidler, he responded, "Folks, this is lower than what the Russians are projecting. We're not going to let Russia beat us. Go back and give me a higher projection that shows the U.S. as a winner." And they did.

The general way of thinking was that the more energy we used the better off we were, and electricity usage was universally accepted as our best barometer of progress. This thinking was behind the industry's "Live Better Electrically" slogan, first presented in 1956 jointly by the utility industry and corporations such as General Electric and Westinghouse. Although primarily intended as a means to sell more electricity and the products that used it, there was a real sense in the promotion that the widespread use of electricity was best for all. The efficiency of the electric appliances of the 1950s left a lot to be desired, however. As the price of electricity rose, so too, did the expense of maintaining the house.[104]

When in 1969 Dave first advanced the notion that we should pursue energy efficiency, he was fiercely opposed. On one memorable occasion, in a meeting in 1969, when Richard Nixon was president, with electric utility executives in the office of Secretary of Interior Rogers Morton, Dave, who was then a staffer in the White House, suggested that the country should start conserving electricity and he was lambasted by industry leaders. They called him a socialist and anti-American to his face in the interior secretary's office. They claimed the use of electricity and economic growth were the equivalent of Siamese twins. The utilities wanted to sell more electricity, as many still do, and many saw increased consumption as fuel for economic growth. But it is capitalism, not socialism, that has driven the change of perspective.

Every business in America wants to achieve greater efficiency of all kinds, and energy efficiency is now widely seen as part of the heart of the success of the American economy. The legislation passed in 2015, sponsored by Senators Rob Portman, Ohio Republican, and Jeanne Shaheen, New Hampshire Democrat, was successfully pitched to meet this desire. The pitch was that it was good for the economy, good for the environment, was part of a plan to bring back jobs lost, would help fix the trade deficit, and make manufacturers more globally competitive. It didn't hurt the bipartisan support that the bill was relatively modest with few governmental mandates.[105]

Of course, increased use of renewable electricity to substitute for natural gas and oil in heating and transportation is a good thing. It is the way we avoid the dangers of climate change, the way we cut costs to consumers, and the way we greatly increase the efficiency of overall energy use. A good estimate is that the U.S. and the world will use roughly 39% less energy in the year 2050 due to this transformation, in great part from efficiency gained from switching to electric cars and heat pumps.[106]

Indeed it is because we need to increase the total market for electricity that using that electricity all the more efficiently becomes more and more important.

Efficiency Is Not the Same as Conservation

Making our homes more energy efficient and buying high-efficiency cars and appliances in no way involves "doing without," or "freezing in the dark," as critics in the past described. Investments in efficiency actually save money as well as energy. Some of us vividly recall the chill in our homes when in response to the energy crisis of the 1970s we were asked to turn our thermostats down. Such voluntary action to save energy is fundamentally different from investments that create efficiency. They are short-term measures, which are helpful to ward off system-wide blackouts in hot summer months due to spikes in demand for air conditioning. Thankfully, most people respond in those crisis situations, and such conservation is important. But in normal circumstances, how hot or cool you want your home should be a matter of individual preferences and values. Seeking efficiency is different; it's all about getting the same service with less energy, and less expense.

Doing so is in everyone's interest. It's in keeping with the capitalist ethic that has driven American prosperity, which induces every business to cut its costs and every consumer to get the best product and service at the best price. Perhaps we need a new slogan, "Conservatives for Conservation," to drive home the fact that energy efficiency is a conservative idea. Greater energy efficiency has been central to Americans' ability to steadily improve our standard of living. Now it is becoming central to protecting our standard of living.[107]

We Know How to Do This

Over the years we have made great progress in using energy more efficiently through many means. The awareness of efficiency opportunities and the dedication to pursuing them have advanced significantly. Some measures have been government mandated, such as building codes that require greater efficiency and a federal law passed in 1975 that required increases in mileage per gallon for motor vehicles, which has led to the most dramatic gains. Before 1975, cars averaged twelve miles per gallon. The 1975 law got them up to around twenty miles per gallon.[108] A new federal mandate, issued by President Obama, builds on this progress, requiring that by 2025 new cars achieve fifty-four miles per gallon.[109] The Energy Policy and Conservation act resulted in a rule effective in 1979 that requires product labels that allow customers to compare the efficiency of air conditioners, refrigerators, and other appliances.[110] Some state and local governments also provide monetary incentives for both households and businesses to make efficiency improvements, and many have issued higher efficiency standards for new buildings.

Efficiency measures have been advanced voluntarily by energy service companies who can evaluate what efficiency investments are cost-effective, do the work and result in net financial benefits from the savings a customer achieves. Even the utilities are informing customers of savings from efficiency and offering rebates for the purchase of energy efficient appliances.[111] These efforts have proven that the cheapest, cleanest, and most reliable source of energy is the energy we avoid using. As the cost of energy increases, the savings from efficiency increase commensurately, and these savings are not just once; we earn them over the life of the equipment.

There's No Better Bargain

A key objection to making efficiency improvements has been that they cost money. Well, some do and some don't, and whatever cost is involved, earns much more over time than the initial investment.

Energy efficiency measures and their costs vary widely, from simple and cheap methods like screwing in a highly-efficient LED light bulb, to installing a heat pump system for heating your home or buying an electric car. The important fact is that you only need to make the investments that save you money. And the opportunities range from 5-50% of your current costs. And, of course, that doesn't factor in the benefits to the environment.

What's more, some efficiency improvements involve no costs at all. Take the example of a white roof that reflects the heat of the sun rather than absorbing it as a dark-colored roof does. The two roofs cost effectively the same amount, but a white roof cuts down on air conditioning costs dramatically. More typically efficiency improvements require some investment such as, better insulation of a building or spending more upfront for a highly efficient refrigerator, air conditioner, heating, or lighting system. In the future, it will mean investing in the most efficient electric car and heat pump. Ultimately, the savings pay the investment back with interest, often in about five years and sometimes in less. After that the consumer is saving big money.[112]

The trick is in the up-front investment that most people don't or can't make. The efficiency service companies that finance efficiency investments are available to large customers, but not to the average consumer for whom the upfront cost is the barrier.

Dave, when he lead the Tennessee Valley Authority in 1978, put in place a program that overcame these barriers to eliminating the 20%

of electricity usage that is pure waste. TVA financed the investments in efficiency and they were repaid out of the savings as part of the electric bill. TVA had 300 "energy doctors" that made house calls. They audited the customers' home and recommended the cost-effective investments to make, helped the customer select a contractor from an approved list, and then inspected the work before paying the contractor.

All the customer had to do was sign a piece of paper. Half a million people did just that and they saved what a large power plant would otherwise have been needed to generate, at less than half the cost.[113]

That model of "on bill" utility financing and "no problem" customer service needs to be a part of every distribution utility in America. If utilities financed solar panels and efficiency, utility bills could be cut while the climate crisis is cured. But that is not going to happen without stronger governmental action. And some states have already set a good example.

A few states have led the way in making greater efficiency central to their energy policies, most notably California, New York, and Oregon. The utility commissions in these states reward efficiency investments by utilities greater than other investments. Even so, financing and help in making it happen are seldom part of the program. Codes for new construction and labeling of energy-using products have improved efficiency and saved people 10-20% on their normal bill. Nevertheless, we still waste enormous quantities of electricity and money.

A leading governmental agency working in this field is the California Energy Commission, which provides funds for public projects promoting energy efficiency and acts as a clearinghouse for commercial and utility projects such as energy audits and rebates for the community as a whole. In Oregon, a large nonprofit agency, the Energy Trust of Oregon, provides similar services. The trend is for more states to provide similar services.[114]

Zero Growth

The U.S. should be aiming for zero growth in consumption for the existing uses of electricity. This goal may seem quixotic, but consider that California has, with an emphasis on efficiency, achieved zero growth in electric consumption for the 2007 to 2014 period. To be sure the recession has helped, but it's happening.[115]

Economic growth is estimated at 2-3% per year. But the U.S. economy is much less energy intensive than in the past. We are more a service economy. And, with more advancements in efficiency of appliances and smart technology we will use less energy.

Home and building energy management systems, smart technology, and advanced metering has provided the consumer and the utility with new tools that will allow them to manage and save energy in homes, buildings, and throughout the electricity infrastructure. Achieving significant energy savings is possible to offset the growth that would otherwise take place in the current use of electricity. That, we can and must achieve.

All states and the federal government must strengthen requirements and incentives for achieving zero growth in the existing uses of electricity as we expand the market to substitute for natural gas and oil.

We can cut electrical usage for lighting by two-thirds as old fixtures wear out and are replaced by LED lighting. Because LED bulbs last ten years and use less than half of the electricity per unit as the most efficient other current systems, LEDs are likely to revolutionize the lighting business. We can reduce electric usage for air conditioning by 25%. The same can be said for refrigerators and smaller electric appliances.

For the transportation sector, electric cars and trains need less than half the energy from oil now used. And the heat pump is much more efficient than gas, propane, or oil heaters.

An All-Electric America will create a much more efficient, lower cost supply for the American consumer.

CHAPTER 5

The 21ˢᵗ Century Electric Utility

The restructuring of the utility industry that is already under way needs to be directed in a focused manner to encourage a great variety of decentralized entities to become the primary generators of renewable electricity.

New York State is leading the way by becoming a more service-oriented utility, focusing more on distribution than generation and helping to create entities that will market energy efficiency, storage, and distributed renewable generation, the microgrids of tomorrow.[116]

The role of today's utilities can remain extremely important. The majority of customers will most likely remain on the grid with the proper incentives. Solar and storage systems don't have to result in customers leaving the grid. A good estimate suggests that an ideal future energy configuration would involve roughly 7% of the total energy generation coming from rooftop solar PV and the remainder to be mostly distributed by utility-scale solar and wind.[117] Even though much of the generation will be utility-scale under this scenario, the old utility structural models and the methods in which we will pay for and transmit energy are changing.

The electricity system will transform from a centrally generated infra-structure to a distributed model, balanced and monitored by the utility with the help of computer engineering and smart-grid technology. The utility's job will include load management, storage, and coordination of the electricity that will now move multidirectionally, from the consumer to the utility and to the smart grid and back. In effect, the utilities will act as the "orchestra conductors," bringing harmony to power coming from their customers' rooftops, microgrids, and distant wind farms.

The utilities must transform from the old per-kilowatt-hour rate-based utility, where a utility makes money based on how much energy they sell, to a more service and value oriented utility. The utility will get paid for providing these services and reimbursement will be based on value of the provided services rather than simply for how much electricity they sell.[118] We may also see a "transactive" system where prices are associated with the electrons that flow through the various devices and where microtransactions take place at substations, storage units, and energy management systems at the home and on the grid.[119]

Additionally, the ideal twenty-first-century utility will have a custom-er-oriented business model.[120] In this model the customer will go to their utility to find the best home energy system, thermostat, storage, solar system, and heat pumps. We can expect the utility to provide financing and installation coordination as well.

The real issue is whether, even as they begin to transform to the role of service provider and deliverer of electricity, rather than generator of it, this group of companies can overcome their history; a history of promotion of large fossil fuel and nuclear power plants and resisting change for cleaner energy. America's electric utilities need to get on a positive course, promoting and providing services that customers want and society needs.

Although some utilities are beginning to see that they must change and adapt, most utilities—consumer-owned and investor-owned—are reacting defensively to the new reality of climate change and the implementation of renewable energy. They require solar customers to pay what the utility considers their share of costs for the grid and limit the size of solar installations. They can be counted on to try to pile on as much cost as possible on agreements with solar customers.[121]

The solar threat is the first seriously disruptive technology that the industry has ever encountered and it comes in the face of resistance to an inevitable sea change in the industry business model. The solar bypass threat is so real that some analysts are warning of "stranded assets" and even the demise of distribution utilities.

The Conductor

The schematic in Figure 2 demonstrates the twenty-first-century utility, the conductor of the electricity distribution center.

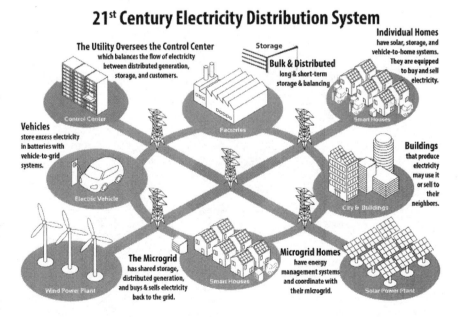

21ˢᵗ Century Electricity Distribution System

The Utility Oversees the Control Center which balances the flow of electricity between distributed generation, storage, and customers.

Storage

Bulk & Distributed long & short-term storage & balancing

Individual Homes have solar, storage, and vehicle-to-home systems. They are equipped to buy and sell electricity.

Vehicles store excess electricity in batteries with vehicle-to-grid systems.

Buildings that produce electricity may use it or sell to their neighbors.

The Microgrid has shared storage, distributed generation, and buys & sells electricity back to the grid.

Microgrid Homes have energy management systems and coordinate with their microgrid.

Control Center
Factories
Smart Houses
Electric Vehicle
City & Buildings
Wind Power Plant
Smart Houses
Solar Power Plant

FIGURE 2: The twenty-first–century-utility will act as the "orchestra conductor," bringing harmony to power coming from their customers' rooftops, microgrids, and distant wind farms. It will manage the storage, distributed generation, and the coordination of the electricity that will now move multidirectionally, from the consumer to the utility to the smart grid and back. Copyright: All Electric America.

"Live Better Electrically"

The U.S. electric power industry, which has an under 50% asset utilization factor and faces a near no-growth future, has an opportunity to do well by doing good. Why not boost business by encouraging the use of heat pumps for heating and cooling homes and speeding the slow-growing market for electric transportation?

Utilities have a rich history of marketing. In 1926 the "Reddy Kilowatt" trademark, seen in Figure 3, was created to help promote electricity and electrical appliances, which even into the 1930's, according to utilities, were not catching on fast enough.[122]

FIGURE 3: The Reddy Kilowatt trademark was created in 1926 by Ashton B. Collins. Collins, with publically owned utilities, promoted the "Reddy Kilowatt Program." Xcel Energy Inc. currently holds the trademark.

Why, for example, don't electric utilities promote electric vehicles and a low-carbon home with the same verve they did the "Live Better Electrically" programs of yesteryear? The advertisement demonstrating their efforts shown in Figure 4 was part of their massive campaign.[123]

FIGURE 4: The "Medallion Homes" campaign was launched in 1957, the goal was roughly one million all-electric homes by 1970. The goal is thought to have been achieved: 300 electric utilities and 180 electrical manufacturers supported the campaign nationwide. Former President Ronald Reagan is seen as the spokesperson for General Electric in this Advertisement.

In 1956, 300 electric utilities, partnering with 180 companies, including industrial giants, Whirlpool, General Electric, and Westinghouse, started a campaign with the goal of creating one million completely electric homes and expanding to rural America.[124] Their main slogan was, "Live Better Electrically." The advertisement shows one of their marketing strategies, a stamp of approval called "Medallion Home," which indicated that the home was completely electric and had been inspected by the local electric utility for meeting the standard modern codes.[125] Even Ronald Reagan was involved before becoming president. He was the spokesperson for General Electric and promoted GE's slogan weekly on television: "Progress is Our Most Important Product."

Electric utilities have a choice: They can continue fighting a losing battle of frustrating customers who want solar panels or other self-supply. Or the utilities can initiate, like Solar City for example, a customer oriented program that might be called "Lend Us Your Roof." Certainly, utility companies can finance and install solar panels as cheaply as anyone. And panels and other energy producing devices don't all need to be on the roofs of buildings. Community solar, wind power, and microgrids can give large buildings and customer groups off-site options as well. And why not combine solar into a utility financial package that also includes efficiency measures, to be repaid out of savings in the bill?

Surely solar delivered directly to the customer should be encouraged and not rejected by arbitrary actions to raise its cost and limit its supply. It's time for utilities to consider the opportunity that PV solar and other decentralized options present.[126] The industry needs to fire its defensive-minded "green eyeshade" folks and replace them with salespeople that act as advocates and offer customers the services they want.

Utilities can do this for a complete energy makeover and prescribe "green doctors" to evaluate a home and promote not just the "All-Electric Home," but the "All-Renewable Home" as well. There is the opportunity of offering their customers a package of solar on the roof and efficiency investments. They can provide better-insulated buildings, heat pumps, lighting, home management systems, and more efficient air conditioning and refrigeration powered by the sun. This will also help utilities have the future opportunity to better manage electricity in the home (manage the load and not just the generation) and even use hot water heaters as storage on the grid.

Dave's experience of forty years reveals that most people are too busy trying to make a living to find the time to invest in energy efficiency. Monetary incentives have achieved only a small fraction of what will save money. Innovative programs are needed to achieve this goal. Utilities should hire "green doctors" who will make "house calls," hire the contractors, inspect the work, and pay for it. It can then require the customer to pay back the investment out of the cost of their energy savings as part of their electricity bill. After the investment is paid back in a few short years, the customer receives the benefit of the saving and a lower energy bill. Such a program will dramatically reduce the demand for electricity over time and thus make it much easier for the utility to promote an all-electric, all-renewable energy supply. In addition, such a program will create new jobs.

Very few utilities are currently this forward thinking, but there are a few. The Vermont utility, Green Mountain Power, lead by innovative and customer-oriented CEO Mary Powell, is implementing programs similar to what we describe. Bill McKibben, Schumann Distinguished Scholar at Middlebury College and cofounder of 350.org, explains how he actually experienced hope when he discovered how a utility supported a complete energy makeover, which reduced a house's carbon footprint by 88% at no net cost.

He explains how, "in the course of several days, coordinated teams of contractors stuffed the house with new insulation, put in a heat pump for the hot water, and installed two air-source heat pumps to warm the home. They also switched all the light bulbs to LEDs and put a small solar array on the slate roof of the garage. The Borkowskis [the home-owners] paid for the improvements, but the utility financed the charges through their electric bill, which fell the very first month." Ultimately, the utility makes money by financing and leasing the heat pumps and solar panels to the customer. When home management systems are implemented they also have the added future benefit of being able to manage the electricity used by the homes.[127]

There are a lot of opportunities forward-looking utilities could seize, even if some require the cooperation of policy makers. Deregulation cut the integrated utility into separate industries—distribution and generation—in the name of competition. It is time for distribution utilities to claim their right to be competitive, too. But for that to happen, it will require that the utilities themselves determine a new course of action that reignites their marketing skills.

Electric power industries must shift their way of thinking to that of taking the offensive by actually promoting products like rooftop solar. They must do away with the familiar refrain about fairness to existing customers when customers want to buy their own panels and "unplug."

Inherent in being competitive is selling something new that some but not all of your customers will want to buy. The utility industry has never applied a strict cost-of-service test before, allowing people to plug in with new appliances, such as air conditioners that require capacity used only a few days in the year and that may actually cost the utility a lot more than the rate paid. There is no cost-of-service test applied to an expansion to a new subdivision. Indeed, the electric power industry has always said "just plug in." We need to do the same with solar, electric vehicles, microgrids, heat pumps, and the new decentralized and smart technologies that a competitive electric distribution industry may produce.

One would have to be blind not to notice the relationship between our petroleum-dependent transportation sector, air quality, and climate change. Electric vehicles offer utilities a chance to come to the rescue. Over time, renewable electricity can virtually eliminate the transportation sector's 71% consumption of U.S. petroleum. The electric car is clearly part of the solution, particularly as more and more of our electricity comes from the sun and wind. On a lifecycle basis, the electric car is cheaper to operate and maintain—equivalent to one dollar per gallon of gas, or better—as opposed to the price of gas at the pump. So as the capital costs come down, it becomes a no-brainer.

The main hurdle for the electric car is its relatively high up front cost and limited range. But suppose utilities could pay for and own the EV batteries for today's electric cars. They would become immediately more affordable. Utilities could justify the investment in part on the basis that the increased off-peak consumption by EVs would add to utility revenue, improve their overall asset utilization and air quality, and reduce greenhouse gas emissions in their service areas. They could even charge the EV customer something for the service and still come out ahead. In the end the auto industry, the EV owner, and utility customers would all come out ahead.

Utilities can alleviate the range problem by installing charging stations, not just downtown, but throughout their service areas. With their long forgotten promotional hat on, they should be able to think of many ways to make it easier to own and operate the electric cars that could add some 30% to its business in the years to come. Moreover, a truly aggressive electric power industry would be proposing electrification of the nation's railroads and encouraging the adoption of the new generation of heat pumps.

A truly competitive electric power industry would promote electric transportation as if its life depended on it—and it might. Over time, utility-owned solar and electricity for transportation will be very cost-effective additions to the system. There are many items that customers and the nation need for this high-energy civilization. Utilities can also promote the transition to electric heat pumps in the way Tennessee Valley Authority and Green Mountain Power are already doing today. They can also be instrumental in the implementation of home management systems that will be able to communicate with the greater grid, the utility, and the devices in the home. Government-supported financing systems would be one way for developing a program. We describe how financing can work in greater detail in Chapter 11: Electric Cars and Heat Pumps for Every Home.

Electric utilities have a choice. They can create a new era of real competition in the industry to supply cleaner electricity and cleaner air by promoting solar power, electric vehicles, and heat pumps, or they can continue playing defense and slowly but surely lose more and more of their customers. It is true that today's electric systems are a big part of our climate problem. They do not have to be. The choice is theirs.

CHAPTER 6

Renewable Energy Costs Less, Not More

The oil, coal, and nuclear folks have dismissed solar and wind power as too expensive for ordinary Americans. They keep saying renewables are expensive, small in size, and only relevant for the future. They ignore the fact that renewables come in any size we need and that their costs keep going down, down, down.

In the meantime, the price of oil, our electric bills, and their "hidden" costs have gone steadily upward. Whether it is electricity rates, costs for gasoline, troops in the Middle East, or medical bills for pollution-related illnesses, not to mention the damages already occurring from climate change, guess who pays for it—no one else but the American people.

Renewables are a better financial bet for the consumer than oil, coal, natural gas, or nuclear power for a host of reasons:

- The total cost to the American consumer is lower over the life of their energy-using equipment.
- The direct cost of the renewable energy is fixed when it is built. There are no fuel costs for solar and wind maintenance and it is thus virtually inflation-proof.
- Renewables are converted to electricity, the price of which is

regulated to reflect costs plus a reasonable profit. This is in contrast to the unregulated price of oil and fossil fuels and the unknown, even higher price of new nuclear power.

- Renewable costs are going down while the price of oil fluctuates with an upward trend.
- The future price of natural gas is most likely to go up.
- The savings in the indirect cost of renewables over coal, oil, natural gas, and nuclear power are profound.
- Some indirect costs include damages from climate change, health expenses, managing the risks of nuclear power, and military commitments—including deployments and even wars to safeguard oil from the Middle East.
- Climate change costs are continuous, incurred each year, as a result of permanent damage from climate change.
- The cost of climate change is expensive and it will cost us more to manage the damage from climate change than to build a 100% renewable energy system.[128]
- Climate policy has been described as "climate insurance," protecting us from irreversible climate change.

The Uncertain Price of Oil

Let's be clear that the price of gasoline has little to do with what it actually costs to produce it. The price is a result of speculators bidding the price up and fear for shortages caused by the combination of terrorism, fear of a Middle East war, and an artificially tight market. Saudi Arabia produces much of its oil for less than five dollars a barrel and has no problem selling it for sixty dollars a barrel. American companies are earning similar windfall profits on U.S. production to the tune of billions of dollars.[129]

There is little or no competition among oil producers. What exists is a well-established cartel called OPEC that is capable of keeping the world on its toes with artificially short rations, which results in high prices for you and windfall profits for them. OPEC's job is more difficult these days because of the fracking boom in America.[130] Even so, cooperation among producers is keeping the price way above the actual production cost of both oil in the Middle East and from existing wells in America. It is still influenced by speculators who fear political embargoes or terrorist acts on the oil fields. Moreover, oil producers everywhere are getting help from the new consumers in China and India where demand for gasoline is growing fast. The "shortages" may not be real and OPEC is fully equipped to keep it that way.[131]

Natural Gas

It is popular these days to say that the price of natural gas is low and therefore is good for the consumer. Never mind that it is as bad a greenhouse gas source as coal and the long-term trend of natural gas prices has been upward. There is no reason to believe that if allowed to increase its amount in the U.S. and for export to Europe that the price won't continue to go up, perhaps sharply.

In the 1960s and early 1970s, when natural gas prices were regulated by the old Federal Power Commission, natural gas was inexpensive. Can you believe wellhead prices were 1.24 dollars (in 2015 dollars) per mcf in 1963? In the 1970s after the prices were no longer regulated, they began its steady climb to the 4.33 per mcf today.[132]

It is useful to note that very little natural gas is sold under long-term contracts. A very cold winter causes shortages and the price shoots up.

The recent excitement about fracking gas can be very deceptive. A fracked well's lifespan is short without continuous investment, likely triggering increased production costs, price increases, and uncertainty. There is no regulation of prices and producers can and do tailor its production to keep the market price profitable for them. No one really knows what the future price of gas will be.[133] Whatever it is, it will be forever uncertain, and in any case, it is too high a price for society to pay since it will keep heating the planet to a point of no return.

The Cost of Renewables

In contrast to the continuing upward spiral of the cost of fossil fuels, the price of solar and wind electricity sold by a utility can be controlled to reflect its actual cost, which is, in fact, on a downward trend. Today, electricity from wind can be generated at or near the existing price of natural gas—in the six cents per kWh range. Coal is usually lower priced. The new coal-fired plants with good controls will cost consumers as much as or more than renewables directly, and will cost much more indirectly from health impacts and by adding to climate change. In fact, today we are beginning to see renewables outcompeting new coal and gas. Significantly, utilities are now choosing to build renewables over fossil fuel plants based on cost alone. And we know what a new nuclear plant in America will cost: more than renewables and the indirect costs will be awesome and forever.[134]

Here's the crucial difference: once the infrastructure is built, the cost of renewables is largely fixed. The fuel, which is most of the cost with oil or coal, is free. As long as the sun rises and the wind keeps blowing, the fuel costs remain the same—zero.

The initial cost of renewables has gone down dramatically in recent years, as anyone who took Economics 101 will know, costs will keep going down as more plants are built and technological innovation allows for more natural "fuel" to be collected and turned into useable energy more efficiently.[135]

We need only look at our hydroelectric plants to see that this is the case. The dams we find in Tennessee, Oregon, Idaho, and Washington State offer examples of states with some of the least-expensive rates and most stable costs in the country. They were expensive to build, but with intelligent financing and investment by the government they provided for low and stable prices for the consumer, resulting in less pollution, increased health, and unbounded growth in the regions.[136]

Fixed Costs of Renewables

There is another basic reason why renewables are a better buy for the consumer. Their costs are pretty well fixed at birth over the long life of the solar panel or windmill. Long-term contracts can assure that the price remains stable and even goes down as adjusted for inflation, which the renewables pretty much avoid because their costs are fixed. This is a crucial distinction for consumers. What you see now, in the terms of price, is what you get for now and the future. As the saying goes, there are few guarantees in life—that the sun will rise and the wind will keep blowing are two of them.

Solar Costs

Solar panels can be used directly to power homes, buildings, and even electric cars. More fundamentally, solar power can be owned by the customer! You can own your own power plant. The costs can't go up— you own it—it's paid for and the fuel is free! There are no moving parts and virtually no maintenance. Sounds like a much better deal than your gas station can offer you![137]

New companies such as Solar City are selling solar installation directly to utility customers providing savings as compared to the utility bill. They bring on a new world of real competition to utilities. Electric companies could fight back and offer solar panels on their customers' premises. Instead most are trying to stop the competition. Both rooftop solar and Big Solar projects of hundreds of megawatts are now a competitive reality. We are also seeing solar plus storage systems being installed, thus bringing even more independence and lower prices.[138]

Wind Costs

Wind power plants reveal dramatically how the energy company lobbies have totally misled the American people. The direct cost of wind power is now competitive with new natural gas, the life-cycle cost of a new coal-fired plant, and in some cases even existing natural gas and coal plants. And wind, like solar, has zero fuel cost. And of course, wind power doesn't carry with it the health bills, global warming dangers, and nuclear war threats of what we are using today.

Wind power also reveals the downward trend. Its costs have tumbled as volume has increased, in contrast to the escalating costs of poisons we are using. It is no wonder that wind power is now favored in France, Germany, and even Texas![139]

Government Incentives and Financing

A renewable power plant (solar or wind) requires little or no labor and fuel to operate. Almost of all its costs are what's required to build it in the first place. That means its total costs are largely determined by the cost of the capital needed to build it—that depends on how it is financed. A plant financed with tax-free debt by a public power agency at 4% interest can cost less than half that of a plant financed in large part by equity that expects to earn 10-20% on the investment. Today there is a crazy quilt of tax credits that are temporary in nature but useful until volume production is achieved.

For the long haul, the best government incentive would be to make the interest on loans for renewables tax free for all renewable power plants and plants used for conversion to hydrogen. At the very least, federal loan guarantees on renewables sold to distribution utilities would carry little risk to the government but would dramatically reduce the cost of renewables and speed their construction.[140]

Indirect Costs

If we examine new power plants, we must look at all of their costs. Burning coal or natural gas creates poisons that contaminate the air, create smog, and bring on global warming. Nuclear plants expose the U.S. to the danger of massive doses of radiation and require major subsidies from the federal government, greater than those inquired by other energy sources, in the form of accident insurance and the security cost of protecting the plants against terrorism. Oil brings with it a laundry list of costs including: contaminating the air, smog, oil spill clean up, global warming, health costs, and military costs in the form of wars, guarding oil at the wells and during transport, and in the loss of lives of our troops.

Good estimates, for example, are that by 2050, switching to renewables would avoid roughly $600 billion per year from air-pollution-related mortalities (roughly 3.6% of the 2014 U.S. GDP) and $265 billion per year in U.S. climate-change costs. Together, this is close to one trillion dollars per year! And air-pollution mortalities and global-warming costs are just a fraction of the total cost of our fossil-fuel addiction.[141] University and business sector economic analyses are warning of catastrophic consequences to the world's economies if we don't reach our climate goals.[142]

If the costs are considered—and they must be if we are to continue life as we now live it—renewables win the competition with a walk. It therefore should be the burden of government policy to mandate the result that is truly the low-cost option because the market prices alone are giving us the wrong answer.[143]

The Need for Regulation of Utilities

If we are able to attain all of our energy needs for our society through electricity from renewables, we will have lower, stable, and predictable prices that will be a safeguard and great benefit for our consumers. The inherent nature of how we price electricity compared to oil and gas make it much better option for the consumer

One thing we do know—competition does not exist among oil and gas companies. These are oligarchies that sell commodities vital to the functioning of our society and, as such, they can charge whatever they want because they have guaranteed customers no matter what. If there are no government controls, the consumer gets ripped off. Greed is directly proportional to grasp. And the price has little or no relation to cost.

The untold story is that the cost to an oil producer in Saudi Arabia or even Texas does not matter. The price of oil shoots up when a country like Iran rattles its sword and speculators get in a bidding contest to avoid shortages. It is much easier for oil producers to pump more slowly if holding back helps to hold us up. Why run the risk of jeopardizing the steady flow of sky-high profits?

Executives for utilities that are regulated follow a rule of thumb—at all costs they avoid raising rates by double-digit amounts at any one time. The reason is obvious. Consumers can't easily adjust to large increases in the cost of an item they can't do without. That's why Americans get angry at the obscene run-up of the price of gasoline, home heating oil, and natural gas, all of which are unregulated. By contrast, electricity prices have increased, but under regulation that increase has been at a modest pace.

Consumers prize certainty and stability in what they pay for electricity and gasoline, items they must keep buying. They are the lifeblood of this high-energy civilization we have created for ourselves. That's a central reason renewables are preferable for consumer protection.

Over the long haul, it is a foregone conclusion that renewable electricity at a regulated price will be cheaper for the consumer than unregulated oil or unregulated natural gas.

Ensuring Efficiency

Getting the price of fuel under control by shifting to renewables is only half the job of how consumers can save big money. Energy efficiency, as we describe in Chapter 4, is an important way for society to help manage the cost of energy. Obtaining energy from renewables is more efficient than from oil, coal, or natural gas and will reduce the amount of energy our society uses. However, we can go even further; we can be more efficient at home as well.

Being efficient is something we inherently understand and buying a car is an example of this in our everyday lives. When you buy a car for $15,000 and finance it with payments of $250 per month, you are also committing yourself to a gasoline bill of $200 per month if you drive 1,200 miles a month at 20 mpg and $3.50 per gallon. If instead you buy a car that gets 40 mpg, you cut your gas bill in half to $100 per month. You save $100 a month.

It doesn't take a MBA from Harvard to figure that one sure way to cut your gas bill is to buy a car with better mileage. The same is true for air conditioners, furnaces, refrigerators, and the like.

The federal government and regulators can help with standards and incentives, but consumers can really help themselves by recognizing that energy—yes, even cheaper renewable energy that feeds your car or appliance over its life may cost about as much as the item itself. Whether you are high or low income, the way to save money without giving up any convenience is to buy the most efficient model. Even if the initial cost is a bit more, that extra cost will pay monetary dividends to you.

What Does the Cost Add Up To?

Mark Jacobson and his partners have a particularly compelling examination of the costs involved in achieving the goal of a 100% renewable energy future.[144] Their study finds that the upfront capital we will need to invest each year is a doable 2% of our Gross Domestic Product (~383 billion dollars per year for 35 years). It is close to only half of what we will pay in 2050 for air-pollution deaths and comparable to what the U.S. will pay for global-warming costs in that same timeframe.[145]

Significantly, China has announced that it will implement a climate plan that will cost 6.6 trillion dollars, utilizing roughly 3% of China's GDP over a 15-year timeframe. China has stated that the plan will not be supported by public finance, but instead expects to attract investors as it adopts new technology.[146] The relative cost of investments Jacobson proposes is similar in scale to China's plan.

This investment is nothing compared to what we put forth in WWII when we contributed 40% of our gross domestic product toward the war and it is less than the 4% of our GDP we currently spend on our military each year.[147] Moreover, unlike our military expenditures, much of the investment will be private.

The final bonus is that by 2050 the U.S. will gain two million long-term jobs, health costs will decrease by roughly $1,500 per year, and the average person will have extra money in their pocket, enjoying energy cost savings of roughly $260 per year.[148] In a United States, where we are losing manufacturing jobs and where our global competition is increasing, it is clear that an investment in renewables is an investment in economic growth, prosperity, and a healthy America.

The surprising truth is that new investments in efficiency and renewable resources will cost consumers less money than the out-of-control market price of oil, the long-term escalating cost of natural gas and coal, and the subsidized and uncertain forever costs of nuclear plants. Once a renewable energy source is built, its costs are fixed and there are NO fuel costs.

Efficiency and renewables are cheaper even on the misleading pricing system we use. If we consider—as we must—the health costs of air pollution and wars, the proliferation and radiation risks of nuclear energy, and the global-warming costs of all fossil fuels, it is a no-brainer. Renewables are the best bet for the consumer.

Part II
Obstacles

CHAPTER 7

The Hidden Truth about Natural Gas

O nly a few short years ago most environmentalists and energy enthusiasts alike were excited by the prospects of natural gas. The promise of a fuel that burns cleaner and pumps out half the carbon dioxide of coal appeared to be the answer to our climate problems.

But the emphasis on reducing coal-burning for electricity generation has obscured the fact that the burning of natural gas for all purposes already contributes almost the same amount of CO_2 emissions in the U.S. as coal burning. In 2014 CO_2 emissions from our total natural gas use were roughly 84% of those from coal and the percentage has grown since then.[149]

The climate debate has focused on emissions from generating electric power, but emissions from residential, commercial, industrial, and transportation sectors are together a larger contributor to global warming than electricity. In fact, 65% of the CO_2 generated from natural gas comes from the residential, commercial, industrial, and transportation sectors of the economy.[150]

The shocking hidden truth is that we actually produce more, most likely much more, greenhouse gasses from our total use of natural gas, for electricity, heating, transportation, and industry, than we do from coal, thus, making natural gas a BIGGER problem than coal.[151]

Natural gas use is rising rapidly. The reality is that, although Btu per Btu, natural gas produces less carbon dioxide, we actually use significantly more natural gas than coal.[152] Natural gas produces enough carbon dioxide alone to cause us to blow past our carbon budget and prevent us from getting to zero greenhouse gas emissions by 2050.[153] Moreover, this doesn't even count the danger from methane released by natural gas drilling and delivery, which makes it an even greater climate threat than coal!

A fundamental problem is that methane emissions are, currently, not being officially measured. The EPA does not require counting and reporting of methane as it does with CO_2. Not counting it does not mean that it is not there damaging the climate.

Natural Gas Is Methane: A Potent Greenhouse Gas

Natural gas is essentially methane. When considering the climate impact of our use of natural gas, we have to consider not only how much carbon dioxide is produced, but how much methane is released as well.

Methane is a different animal than carbon dioxide. It is considered a potent greenhouse gas and is the second largest warming agent next to CO_2. Its impact is greater than is realized. Methane is actually much more dangerous than CO_2 pound per pound. It is 120 times more potent when released, 84 times during the first twenty years after release, and 34 times during the first one hundred years.[154]

The next twenty to thirty years, when methane is the most potent, are critical. It is in these next years when we are most vulnerable to irreversible climate damage if emissions are not reduced.[155] Recently, researchers have begun stressing that reducing shorter-lived warming agents, such as methane, is a particularly effective way of controlling peak warming temperatures and thus limiting near-term impacts of the most disastrous consequences of global warming such as complete glacial melting and the danger of reaching catastrophic tipping points.[156]

If we consider methane's warming potential over a twenty-year period, methane actually accounts for at least 26% of the greenhouse gas (CO_2eq) emissions.[157] These near-term risks make addressing methane a priority.

Methane Leakage:
How Much Is Actually Leaking?

According to a 2015 Environmental Defense Fund (EDF) report, "... the 20-year climate impact of methane escaping from oil and gas operations worldwide has the same near-term climate impact as emissions of 40 percent of total global coal combustion. Using the 100-year metric, oil and gas methane emissions would rank as the world's seventh-largest emitter, coming in just under Russia." Without action, the leaks are expected to grow 23% by 2030.[158]

The amount of methane that actually leaked from the total life cycle of natural gas—that is the drilling, production, storage, distribution, and use of natural gas—is still being hotly debated. Research is under way to try to determine the full extent of leaking.

There are two types of methods for determining methane leakage, "bottom up" and "top down." "Bottom up" studies measure leakage at the emissions source, such as at the wellhead, storage location, or pipe. Total U.S. or world estimates are then extrapolated from data taken at a particular location. "Top down" studies are aerial measurements taken from an aircraft, satellite or tower.[159]

Almost all "bottom up" studies in the United States on conventional drilling have found methane leakage to be roughly 1.5-4% of natural gas's entire life cycle. Most "bottom up" studies have found that the estimated total life-cycle emissions from unconventional drilling (e.g., fracking) are between 2-6%.[160] Recent "top down" studies have shown higher leakage rates of roughly 1.5-17%.[161] Organizations such as the IPCC, after analyzing the different studies, are finding that emission estimates are generally falling between 2-5%.[162]

Using the "bottom up" approach, the EPA developed emission estimates of 1.5%. Recent "top down" analyses, however, have put the EPA estimates into serious question. Stanford-lead and Harvard-lead studies have further put the EPA's numbers into question. Analyzing the data from the different studies on methane leakage, the researchers found that methane emission rates have been underestimated by 1.5-2 times (roughly 50%-100%).[163] A recent August 2015 study identified a potential source of the EPA's low estimate discrepancy as coming from facilities that collect natural gas from multiple wells. It is estimated that these facilities are losing about eight times more than estimated by the EPA.[164] The EPA's estimates fall within the lowest of all "bottom up" and "top down" estimates. Acknowledging the shakiness of their numbers, the EPA said in its greenhouse gas inventory report that it is aware of the new studies and ". . . has engaged with researchers on how remote sensing, ambient measurement, and inverse modeling techniques for greenhouse gas emissions could assist in improving the understanding of inventory estimates."[165]

There are also serious concerns about methane leakage in distribution systems, particularly in older cities. A report by U.S. Senator Edward Markey (D-MA) stated that, "gas distribution companies in 2011 reported releasing 69 billion cubic feet of natural gas to the atmosphere, almost enough to meet the state of Maine's gas needs for a year and equal to the annual carbon dioxide emissions of about six million automobiles." In some cases leaks are extreme and dangerous, including pipeline explosions in locations including, Missouri, Pennsylvania, New York, California, and New Jersey.[166] The Markey report stated that, "From 2002 to 2012, almost 800 significant incidents on gas distribution pipelines, including several hundred explosions, killed 116 people, injured 465 others, and caused more than $800 million in property damage." Most "bottom up" studies have estimated that the distribution system leaks in a range of roughly 0.4-2.5%.[167] However, a few studies have recently found that leaking from distribution pipelines within cities, particularly those with aging infrastructure in the northeast, is higher than previously thought. A 2014 study found leakage rates at 2.1-3.3%, larger than what most industry reports have determined.[168] Another study, however, found lower rates, reporting that leakage rates vary by region and age of the system. They conclude that, based on the areas they analyzed, their emissions estimates were actually 36% to 70% less than the EPA 2011 inventory estimates for distribution and hypothesize that this reduction is from upgrades and modernization of the infrastructure.

Nonetheless, as indicated by the Markey Report, our distribution systems are leaky and we cannot deny that the extent of methane leakage in our distribution systems is quite large in many areas throughout the United States. Overall the true extent of distribution system leaking is not well known or well documented. Distribution system leaking should not be ignored, not only because it poses serious safety issues, but because it is also an important contributor to greenhouse gas emissions.[169]

Total Natural Gas GHG Emissions— Greater than Coal

Methane leakage makes natural gas a dangerous greenhouse gas agent. The surprising result is that we are almost certainly producing significantly more greenhouse gas emissions from natural gas than from coal.

A comparison of the emissions from coal-burning and natural gas at different leakage rates, as shown in Figure 5 reveals the magnitude of the problem. A conservative estimate assumes that the leakage rate is somewhere between 1.5-5%. Our calculations show that, at the EPA's current estimated leakage rate of 1.5%, greenhouse gas emissions by coal and natural gas are virtually the same. But considering that the breadth of studies indicates that this estimate is too low, the implications of methane leakage are huge.

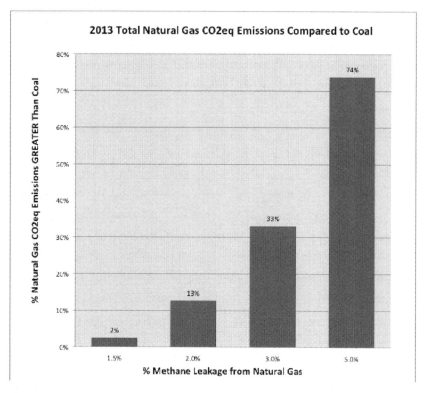

FIGURE 5: The chart demonstrates the percent of natural gas CO_2eq emissions GREATER than coal when considering total natural gas and coal use for electricity, residential, commercial, and industrial uses. Carbon dioxide equivalents (CO_2eq) include CO_2 + methane emissions. Copyright: All-Electric America

Natural gas is a much greater threat than coal and we are heating the earth at a faster rate than we think. What is particularly concerning is that even at the midrange of the high and low of conservative estimates of methane leakage natural gas's emissions are 33% worse than coal. At the higher conservative range we hit natural gas emissions of roughly 74% greater than those from coal.[170]

The Siren Call

Determining how much methane is actually being released from our natural gas use will give us a more realistic picture of where we stand with regards to climate change. In addition to considering the massive methane leaks from electricity, heating, transportation, and industry we also need to count methane emissions from agricultural land use and waste.

According to the EPA 2015 Inventory, only 25% of methane emissions are from natural gas systems. The other 75% of methane we produce comes from animal agriculture, landfills, coal mining, petroleum systems, wastewater, forestry, rice agriculture, and other miscellaneous activities. It is necessary to reduce, manage, and eliminate methane release from all of these sources to most effectively limit the effects of climate change.

The White House, in March 2014, acknowledging the need to have more effective regulation, documentation, and control of methane emissions, proposed a "climate action plan strategy to reduce methane emissions." The plan has set a goal to reduce methane leakage by 40-45% from 2012 levels by 2025.[171]

As part of the implementation of this plan, the Obama administration has directed the EPA to draft new rules, under the Clean Air Act, for reporting and reducing methane and volatile organic compound (VOC) emission in the "Oil and Natural Gas Sector." However, it is not clear how the U.S. will actually achieve the climate-action-plan goals. The rule will only apply to methane emissions from new and redeveloped pipes and wells (effective August 23, 2011) and not preexisting infrastructure.[172] This is concerning considering it has been estimated that the existing sources (e.g., the infrastructure in place prior to 2011) will be responsible for up to 90% of the leakage problem through the year 2018. Therefore, as it stands, this act and the new rule is unlikely to significantly impact the problem.[173]

As of November 2015 there were no requirements for reporting emissions, just a call for voluntary action. Once the revised rules are effective, any infrastructure before August 23, 2011, will remain only subject to voluntary efforts.[174] In addition to this lack of reporting in the oil and gas industries, agricultural emissions and landfill venting are not regulated even though they account for the largest fraction of the methane emissions. Most importantly, even if reporting is achieved, total natural gas emissions will still remain on par with those of coal.

The calls for switching our electricity infrastructure from coal to natural gas are misinformed and misguided. A number of studies have evaluated whether natural gas might be a good interim fuel (e.g., a bridge) to replace coal and found no real benefit. Tom Wigley, professor at the University of Adelaide and fellow of the American Association for the Advancement of Science, explains that, "even with zero leakage from gas production . . . the cooling that eventually arises from the coal-to-gas transition is only a few tenths of a degree Celsius. Using climate amelioration as an argument for the transition is, at best, a very weak argument."[175] The resounding conclusions of these studies are that there is no actual value.

The siren call of cheap natural gas is powerful. The misrepresentation that natural gas burns cleaner than coal is lulling the public and policy makers into believing that its use is better than coal and oil and viable strategy for preventing climate change.

The recent advocacy for the U.S. to export natural gas in liquefied form is just another example of leaders turning a blind eye to the problem. They are ignoring studies which have shown that the process of liquefying and then gasifying the fuel is responsible for particular egregious leakage and indeed produces greater greenhouse gas emissions than oil.[176]

Unfortunately, our current "All of the Above" policies are sending us in the wrong direction. They are further encouraging the "fracking boom" and promoting exports. And the boom is spreading around the world. In the fall of 2014 the Mexican government announced that it intends to develop fracking, as have those of Ukraine and Lithuania. We can only expect the extraction of natural gas to spread globally as more deposits are found.

Our addiction to natural gas is hard to give up. The industry, the federal government and many state governments, as well as consumers, have enjoyed large benefits from the fracking boom and the resulting cheap extraction of natural gas. What haven't been factored into our policy decisions are the climate and health costs of a fossil fuel that leaks methane and produces enough carbon dioxide alone to blow our carbon budget before the year 2050 and lead us on a path to climate hell.

CHAPTER 8

Keep It
in the Ground

The world's energy consumption is growing at a breakneck pace. If we continue increasing our use of fossil fuels, the concentration of greenhouse gases will lead to warming that will severely curtail our current way of life. Only the most zealous of deniers any longer dispute this. Indeed the scientists tell us that we must start reducing greenhouse gas emissions at once and replace all fossil fuels in thirty-five years.

The United Nation's International Panel of Climate Control has determined that the maximum amount of greenhouse gases the atmosphere can safely hold is 790 gigatons of carbon (GtC) if we are to prevent raising the Earth's temperature more than 2°C above preindustrial temperatures.[177] The United States and 114 other parties agreed during the United Nations 2009 Copenhagen Conference that a 2°C increase is the safe limit.[178] Above this level we risk serious disruption to the environment and society.[179] The 2015 United Nations Climate Change Conference (COP 21) was scheduled in Paris with the intention of developing a legally binding agreements on this safe limit.[180] [Note: The 790 gigatons of carbon accounts for the impact from carbon dioxide as well as other warming agents such as methane. It is derived from carbon dioxide equivalents (CO_2eq), the concentration of CO_2 that would cause the same level of warming as a given type and concentration of different greenhouse gases.][181]

The people on Earth already emitted roughly 520 GtC by the year 2012, so humankind has only 270 GtC left to safely add after the year 2012.[182] If we continue adding greenhouse gases to the atmosphere at our current rate, we are set to blow this budget by roughly 2033.[183] The surest way of staying within our greenhouse gas budget is to get on a path NOW to reach zero greenhouse gas emissions by midcentury as we will describe later in this chapter.[184]

Many studies that have evaluated our greenhouse gas budget use a budget that considers only carbon dioxide emissions. In this book we use a budget that includes other greenhouse gases as well.[185] It is crucial that we consider the accumulation of all greenhouse gases, particularly methane, when creating a policy to effectively control greenhouse gases. Without considering all greenhouse gases we do not get a realistic picture of how fast we are in fact warming the earth and the appropriate measures we have to take to reach our climate goals.[186]

We aren't making enough progress in the developed world in slowing the increase of emissions, let alone reducing them; making matters worse, the developing world is rapidly increasing them. Projections are that worldwide greenhouse gas emissions will explode by 70% from 2010 to 2100, with developing nations' emissions more than doubling in places such as the Middle East and Africa.[187] This is due to both population growth and rapid modernization. If major programs to develop renewable sources are not undertaken by these nations, it is projected that almost 80% of the energy they consume will come from coal, oil, and natural gas, mirroring the energy use of the developed nations through 2050. Even if developed nations begin to achieve near-zero growth in emissions from energy efficiency measures and increased renewables, this is not enough to counteract the increase in emissions from growth in the other nations of the world.[188]

The developing world's portion of global emissions is already roughly 60% of global emissions today, and we can expect it to increase.[189] Moreover, as of now, 1.2 billion people are living without electricity in these countries. If they are connected to the grid, they could add hugely to the world production of electricity in the twenty-first century.[190]

As the developing world modernizes, world electricity production is expected to increase by 91% and the number of vehicles is expected to grow by roughly 5 times in China and India and 3.3 times in other rapidly growing and developing nations between 2010 and 2050.[191]

How can we argue, though, that the developing world shouldn't build modern cities and that its citizens shouldn't enjoy the same luxuries— the same cars, appliances, and gadgets—that we in the developed world love so much?

The primary solution is that, as these societies modernize, they build a renewable energy infrastructure, and that the developed world does so as well. It would also help if America provided an example of controlled material affluence by being less wasteful, actually increasing product quality and reduce our consumption of material goods.

The biggest obstacles to us moving forward with this vital work are the inertia and reluctance to change a system that has plentiful resources of fossil fuels and the fact that companies and governments are making enormous sums of money selling them. Good estimates are that the world still has proven reserves from oil, natural gas, and coal that would produce 1,200 GtC (4,400 gigatons of carbon dioxide).[192] This quantity, according to a remaining 270 GtC (990 gigatons carbon dioxide) greenhouse gas budget, is four times greater than the amount that, if released, will bring disaster.[193] Fossil fuel corporations, both private and government owned, would like to burn it all. Avoiding disaster requires keeping 75% of our fossil fuel reserves in the ground.[194]

This book documents the truth that all the fossil fuels and nuclear power are poisons that we must stop using. The case against coal, natural gas, and oil as dirty carbon intensive fuels is widely under-stood. But the public has been led to believe by the EPA, the Obama administration, and the natural gas industry that natural gas is a useful "transition" fuel despite the concerns about fracking and the impact of methane. Calls by some to renew our commitment to nuclear power also ignores its dangers and its high cost. We address both concerns in this book, but first we examine the reasons we must keep all of them in the ground.

Staying Within Our Budget

The gap between where we must be and where we are has become crit-ical.[195] United Nations Under-Secretary-General, Achim Steiner has sounded a call for political action, warning that, the current pace of action is insufficient to reduce emission levels to a level consistent with the "2°C target."[196]

Without changing course, world emissions have been projected to increase at a rate of roughly 2% per year.[197] No matter how long the delay, the total amount of greenhouse gases we can safely put into the air is the same.

Figure 6 demonstrates how on our current path, using the IPCC 270 GtC budget, we will blow our greenhouse gas budget sometime around 2033.[198] Figure 6 demonstrates that to reach our climate goals, green-house gas emissions should peak by 2015 and decline gradually by 3% per year from the 2015 levels, reaching zero GHG emissions by midcen-tury.[199] The later we peak, the more drastic the reductions we will have to make over a shorter period of time.[200] As seen in Figure 7, peaking in the year 2030 compared to 2015 results in severe 20% per year green-house gas emissions reductions needed to stay within the remaining greenhouse gas budget. Peaking in 2015 allows the world to reduce emissions at an easier and doable rate of 3% per year.

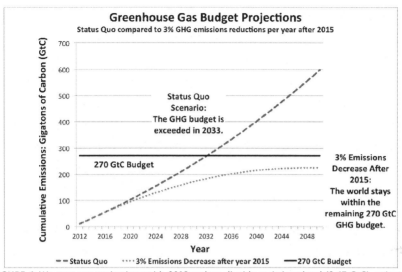

FIGURE 6: We start our projections with 2012 carbon dioxide emissions level (9.67 GtC) estimates from the USDOE Global Carbon data. The carbon emissions include burning fossil fuels, gas flaring, and cement use and 0.9 GtC for land use. 270 gigatons of carbon (GtC), obtained from the IPCC, is the carbon equivalent greenhouse gas budget based on carbon dioxide and other greenhouse gases such as methane. 270 GtC is the greenhouse gas budget remaining from the year 2012. Copyright: All-Electric America

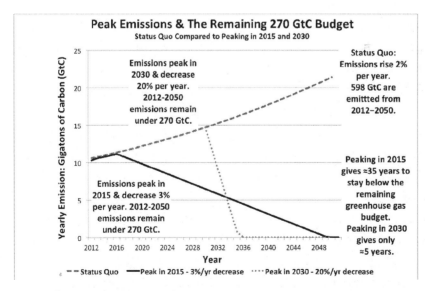

FIGURE 7: Peaking emissions in the year 2030 results in drastic 20% per year greenhouse gas emissions reductions needed to stay under the remaining 270 GtC greenhouse gas budget. Peaking in 2015 allows the world to more slowly reduce emissions at an easier and doable rate of 3% per year. Copyright: All-Electric America

Greenhouse-Gas-Budget-Based Policy

World leaders have made pledges at United Nations Summits, but a pledge is not a program.[201] Despite having a 2°C ceiling, there is a lack of understanding and clarity about what we must do to achieve this goal. Policy makers, environmentalist, climate experts, and academics are all guilty of giving only lip service to the goal of staying below 2°C. However, the proposed plans today have little hope of reaching this objective. Too few are advocating what is needed, a 3% per year reduction in natural gas and oil consumption in the U.S., which is required if we are to do what the scientists say we must; bring these emissions down toward zero by 2050.

There is a call for carbon capture and storage as well as geoengineering (scrubbing carbon out of the air) as a way of staying within the 2°C limit and thus giving us until the end of the century to reach zero emissions. However, carbon capture of coal has encountered a myriad number of technical problems and costs are presently so high that solar and wind energy is now a bargain in comparison. The technology won't be ready anytime soon. Finding safe places underground to store super-huge quantities of carbon and be assured that it will stay put for decades assumes a knowledge of geology we do not possess.[202]

We don't need to rely on expensive and unproven carbon capture and the untested technology of scrubbing carbon out of the air. We have technology, in the form of solar and wind energy, readily available today to solve all of our climate problems. Relying on untried and expensive ideas instead of transitioning to renewable energy available today, and a competitive commercial product, is gambling with our future.[203] The surest and proven way for successful, CHEAP, and clean energy is renewable energy and storage.

"Keep It in the Ground"

A crucial fact that goes unstated by the politicians and most environmental spokespeople is that natural gas and oil emit roughly 70% of the carbon dioxide in America. Coal accounts for only 30%.[204] Yet, the proposed rules being advanced by the EPA target coal and largely ignore natural gas and oil. Actually, the situation is only going to get worse because the fracking boom is resulting in an increase in natural gas and oil production.

There is lots of green talk going on and the beginnings of a "green boom" to produce electricity. We are making some progress in greening the electric power sector due to the fact that solar and wind are comparable in price and are mandated to a varying extent in over thirty states. But that is for just the 30% of the carbon dioxide emissions from the electricity sector. In the 70% caused by oil and natural gas for heating, transportation, and industrial processes we find nearly all politicians and the majority of the general media saying it is good news that we are rapidly increasing oil and gas production with no serious constraint on consumption. And this ratio does not even count the methane release by natural gas.

If one reads the financial pages of "American Press" there is nothing but joy over the oil and fracking boom. According to the United States Energy Information Administration and the natural-gas spokespersons who appear on television every evening, America has become number one in natural gas and oil production.[205]

We are making some progress in energy efficiency standards. President Obama enacted a mandate for 54.5 miles per gallon by 2025.[206] Though this is progress, it is not enough. Dave, as the congressional staffer who proposed and helped enact the CAFE standards in 1975, knows the value of increased efficiency in cars and trucks, but efficiency at most is slowing down the growth of greenhouse gas emissions. Without well-defined requirements and greater incentives for electric vehicles, electric railroads, and electric heating, greenhouse gas emissions have no chance to be reduced 3% year, year in and year out.

The contradiction between the good news about the increased

production of oil and gas in America must be confronted with the fact that this "good news" is really very "BAD NEWS." We can be sure that if we keep on drilling and fracking what we produce will be consumed. It is the last hurrah of an industry that is emitting a unique form of poison that may well end life on Earth as we know it.

Since we have already discovered 75% more fossil fuel than we can safely burn in the future, common sense would suggest that we bring an orderly halt to drilling ever more oil, natural gas, and coal. The climate change threat, and from that perspective, the threat of business as usual, is that continued production and use of oil and natural gas is truly alarming. We need to "keep it in the ground."[207]

A vivid example of why, "keep it in the ground" needs to become a central objective of those concerned about climate is the controversy over the Keystone XL Pipeline proposal to bring tar-sand oil to America. The pipeline presented worrisome problems on which the debate focused, but the fundamental problem was that we did not keep the oil sands in the ground. Once it is dug up it is going to be burned somewhere. The fight is going to be won or lost at the place the gas or oil is buried. It needs to stay buried. And while the environmental leaders make the XL Pipeline their most vocal issue, new drilling remains a possibility in the Arctic and the president is offering more new leases in the offshore Atlantic.[208]

The media is hyping both the "green revolution" and the "brown surge" without pointing out that if the "brown surge" wins over "green revolution" the planet and humanity lose. This failure to connect the dots is in part due to the "All of the Above" policy. This policy does not realize that we cannot overcome the climate challenge with both a "green revolution" and a "brown surge" in America.

We are not naïve enough to believe that governments can just stop future drilling in one fell swoop. But, we believe it can and must happen step-by-step to reach our ultimate goals.

President Obama argues that congress is blocking his climate agenda and thus he is forced to use his executive authority. A dramatic and useful first step for our current president, and indeed any thereafter, would be to use their executive authority to order that there will be no new leases issued for drilling or mining on federal lands. Such an action would dramatically raise the issue for public discussion. The action could and should lead to similar actions by state governments, as well as inspiring citizens concerned about the climate to stop any more drilling on their private lands.

In contrast to inaction by politicians, citizen campaigns, such as "Keep it in the Ground" and "Divest from Fossil Fuels" lead by Alan Rusbridger, former editor-in chief of the Guardian and Bill McKibben, founder of 350.org and Schumann Distinguished Scholar at Middlebury College, are urging the world to take a stand and divest from fossil fuels.

Fracking causes water contamination and even earthquakes. Offshore drilling causes oil spills and drilling in the Arctic ruins pristine environments that should be left alone. True enough, but the debate needs to focus on one essential fundamental point. The continued burning of coal, oil, and natural gas is perhaps, next to the possibility of nuclear war, the greatest threat that all of humanity has ever faced!

"Keep it in the ground" is the heart of what must happen if the climate challenge is to be overcome. We must begin now.

CHAPTER 9

Nuclear Energy: Spare Us the Cure

The atomic age started when the U.S. dropped "The Bomb" on Japan. Let us hope and pray that it won't end with another atomic explosion.

This reminder is necessary because it is important to know what sparked the development of atomic energy. It was a nationwide guilt trip. President Truman's reaction to his atomic bombing of Japan was to say we needed to make "a blessing of this terrible event."

The initial promises about atomic energy go way beyond the oft-quoted phrase "too cheap to meter." Robert Hutchins was the president of the University of Chicago when much of the research for the Manhattan Project took place. He said atomic energy would produce "heat so plentiful it will even be used to melt snow as it falls."[209]

A science writer for Scripps Howard newspapers wrote that atomic fuel would enable "the privately owned airplanes that were suitable only for cross-country hopping to fly across the Atlantic." There was talk of pill-sized pellets of atomic energy that would run a car for a year.[210]

In short, the nation was led to believe that atomic energy would usher in the "Age of Plenty." And there was not even a hint that atomic energy itself might pose any safety risks. Indeed President Eisenhower's "Atoms For Peace" speech to the United Nations on December 8, 1953, kicked off a sixty-nation tour to sell the so-called peaceful atom all over the world.[211]

It is useful to recall the euphoria in which the nation viewed atomic energy at least in the first twenty years. Unfortunately, many scientists, and indeed large segments of the public, developed what we call a religious belief in favor of atomic power. People thought that surely the folks smart enough to make the bomb could perform these peaceful miracles. That may be a basic reason why there are many people who still can't accept the fact that the peaceful atom was all a dream.

No Such Thing as a Peaceful Atom

The harsh truth is there is no such thing as a peaceful atom. The enriched uranium for an atomic nuclear plant only needs further enrichment to make a bomb and plutonium serves both purposes. Our dispute with Iran illustrates the point vividly. There were ongoing tensions when Iran insisted on their right to enrich uranium to fuel the atomic power plants promised to them under the "Atoms for Peace" treaty they signed.[212]

The "Atoms for Peace" program promised any nation that signed the treaty has the right to build atomic power plants with the condition that they do not make nuclear weapons. Unfortunately the fuel they need to power a power plant is everything they need to make a bomb. And with nations that we don't trust to keep their word we find that we promised the ingredients of war, not peace.

American attempts to keep other nations and terrorists from obtaining atomic bombs have little credibility as long as America continues to promote atomic power plants for itself. The danger is not only a nuclear Iran but the availability of plutonium from reprocessing spent fuel and creating more plutonium—the fuel for the bombs. If even a small amount of plutonium gets into hands of groups like Al Qaeda and ISIS, for example, we could expect a disaster in a big city that would make 9/11 look like a traffic accident.[213] And the only sure way to prevent it is for America, home of the so-called "Peaceful Atom," to declare atomic energy the very dangerous failure that it is.

The atomic power proponents want the public to forget that the "Age of Plenty" they envisioned depended on building atomic breeder reactors. The current light-water reactors were never supposed to be the end result—they were just the first step toward the breeder reactors, which were promised to the public as a "Pacific Ocean of energy," a "too large to measure" source of energy. The idea was to take the spent fuel from the light-water reactors and reprocess it to extract the plutonium to fuel the breeders and then bury the wastes in the ground somewhere.

The problem is that the breeder reactors failed both economically and operationally here in the United States, in France, and elsewhere. Over $100 billion, in 1996 dollars, has been spent by nations trying to commercialize breeder reactors that use reprocessed plutonium. All they've created are nuclear proliferation concerns, with no commercial success.[214]

To emphasize the truth again, there is no such thing as the peaceful atom. We cannot, with a straight face, pursue atomic programs at home, with all our bombs and nuclear power plants, and then persuade other nations not to make a bomb. Nuclear power plants themselves are inherently dangerous and are huge threats to domestic security. It is time to recognize that atomic energy is a deadly hazard to humanity. "Atoms for Peace" is just another name for "Atoms for War."

Now that we have a factual understanding of the small size of the atomic energy potential, know its dangers, and have economical alternatives to climate-changing fossil fuels such as solar and wind, it is time to act on that knowledge. The awesome impacts of Three Mile Island, Chernobyl, and Fukushima should eliminate the myth that atomic power is safe. But the record is far more extensive and conclusive than even those awesome events.

The Great Failure

Nothing gives a greater understanding of what a failure nuclear energy has been than having firsthand experience with managing nuclear energy, an experience, in fact, that one of the authors of this book has had. Dave Freeman has been an active participant in the electric power industry over the last forty years. He has been a regulator, a policymaker, and the CEO of a number of large public power systems that operated atomic power plants. He knows of no failure as clear and dramatic as the failure of atomic power on both public safety and economic grounds.

Historically, nuclear power has built up high expectations from the American public but has followed up with only a series of failures. Fifty years after the first commercial nuclear reactor began service in Shippingport, Pennsylvania, in 1957, the nuclear industry has failed on all counts:[215]

- Failed to find a safe place to store its wastes
- Achieved the largest cost overruns in utility history
- Failed to perfect any of the fundamentally safer reactor concepts
- Failed to develop the breeder reactor that would make nuclear a large source of usable energy
- Failed so miserably in the marketplace that no nuclear plants have come online in the US for fifteen years, and only two have been built in thirty years
- Failed to be safe or economically viable enough to obtain its own insurance after more than five decades and billions of dollars in research

Yet the advocates of nuclear power have launched a new initiative that fails even to address the failures of the past. Instead, they stress that nuclear energy is carbon free and is the sure cure for global warming.[216]

Nuclear power demands perfection in its manufacture, its operations, and its maintenance. It also demands instantaneous reaction to trouble. Given those challenges, if we build enough of them, America is going to have a lot of trouble.

Some of the most vocal advocates for dealing with climate change are supporting atomic power; they are dead wrong. Atomic power is a clear and present danger to life as we know it. No one with real knowledge of the dangers posed by atomic energy could support this radioactive menace as the cure for anything. Maybe thirty years ago, but not today when solar, wind, and storage are commercial, can be built faster, and over their life provide lower-cost electricity than atomic power.

Spare us the cure. All radiation, including the routine emissions referred to by the pseudo-magical term "low-level" radiation, can cause cancer. There is no threshold below which radiation emitted routinely by nuclear power plants is absolutely safe. In addition, radioactive tritium has leaked into the groundwater at twelve operating nuclear reactors. Radioactive wastes are dumped into waste dumps not licensed for that purpose.[217]

It is important that the large number of younger Americans who have not heard about nuclear dangers and failures are told the truth. The truth will not come from the federal government, which is pronuclear, but the facts are out there.

Dangers: A Closer Look

Let's examine some of the very real and specific dangers of nuclear energy. The nuclear reactor's potential for massive destruction is an immediate danger. Nuclear plants are vulnerable to accidents, Fukushima vividly reminds us of that terrifying fact. The younger generations don't remember Chernobyl and Three Mile Island, but those older than forty-five should. Those were terrifying days for the entire world and the lessons learned should never be forgotten.

Nuclear power plants have now become a prime target for terrorists. Terrorists have the opportunity to hit the U.S. with the radioactive force of a massive atomic bomb without even having to smuggle weapons across our borders. Make no mistake, these nuclear plants, especially the pools full of radioactive fuel, while well fortified and policed, can be penetrated by an airplane or missile attack or a well-planned terrorist attack. Such events threaten the damage of a bomb. It would take only one.

The atomic power cycle provides the means and excuse for politically unstable and unfriendly nations, as well as terrorists, to build an atomic bomb.

Nuclear power generation creates "spent fuel" that remains highly radioactive for hundreds, even thousands of years. It won't go away; there is no safe grave or burial site. This is the "forever" problem. Not only have individual states passed laws about not having radioactive waste deposited within their boundaries, but several have passed laws that assure that radioactive waste will not even be transported across their state line! There must be something wrong with these waste products if no one wants them in their backyards. This is an immoral radioactive legacy we leave our descendants. Alone, this is enough reason to stop making more nuclear waste.

The mining of uranium exposes miners to lung disease and leaves behind radioactive wastes in the form of mill tailings that remain dangerous for hundreds of thousands of years.

It is not necessary to agree on the size of the nuclear risk. Even the nuclear industry must agree that it is much greater than zero! Why take such a risk when there are benign alternatives? No sane nation in a war with terrorists would consider building more and more nuclear plants. The Obama administration, Congress, and oddly some in the environmental movement are trying to do just that. It is bad enough that we are stuck with old, decaying nuclear power plants that must be retired in an orderly fashion. New nuclear facilities will just increase the risks of accidents and provide more targets for deliberate meltdown by terrorists.

A Cost Too High

Now for the costs. As mentioned earlier, despite all these safety risks, nuclear power also has failed economics. Only a very few atomic plants have been ordered since 1973, and this market failure is not a mystery. Nuclear was hyped in the 1960s as being "too cheap to meter" and ended up since the 1980s as "to expensive to build."

Nuclear power is a poor economic risk also because no private insurance is available for an accident that causes billions of dollars of damage. No utility is willing to take the risk without such insurance. The cost is too high. Without the subsidy of federal insurance, the nuclear option is dead. The federal government—meaning the American taxpayer—takes the risk. Federal loan guarantees and guaranteed funding by utility consumers while the plant is being built add up to a giant subsidy.

Subsidizing the risk inherent in nuclear power is only the first basic assist that the federal government gives the nuclear industry. Between 1948 and 1998, Congress approved $66 billion on nuclear power research and development, and the subsidy funding gets larger, not smaller. The Energy Policy Act of 2005 approved over $4 billion in tax breaks. These amounts dwarf the puny incentives for solar energy.[218]

Of growing proportions are the costs of protecting atomic power plants with aerial surveillance and added homeland security. The nuclear waste problem is largely viewed as a safety concern. However, if there is radioactive trash that will be lethal for many thousands of years, there is a cost to try to keep it reasonably safe. These costs will go on and on and on. They are large and pretty much incalculable. We can be sure of only one thing, our kids and grandkids will pay the bill.

Straight Talk: The Choice Is Clear

The American people, when asked to compare nuclear with solar and wind, know better than the so-called experts. No one in his or her right mind could prefer a radioactive factory to one that harnesses the sun and wind! As a substitute for all fossil fuels and nuclear energy, the sun and wind can replace these poisons with inexhaustible fuel. By contrast, it must be noted that the uranium supply in the U.S. is a relatively small source of energy.

Renewables pose none of the hazards of atomic energy, are far more plentiful, and are much lower in cost to the consumers who ultimately pay all the bills. Rather than granting massive subsidies to resurrect a technology that is inherently dangerous and a failure, let's put American money and ingenuity into advancing a safe and superabundant alternative with costs that are inflation proof (the sun goes up, but not its price).

Even the nuclear utilities are facing the truth. The executive vice president of Exelon, the nation's largest nuclear utility said in the spring of 2015, that when speaking of nuclear power, "we think the more economic alternative right now is renewables, and storage and energy efficiency."[219] We agree.

Part III
Opportunities

CHAPTER 10

Hydrogen Is
the Hopeful Future

Hydrogen is the most plentiful element on Earth. Water is two parts hydrogen and one part oxygen (H_2O). When hydrogen burns, it produces only water vapor—no pollution. But it is not found in pure form in nature. It must be separated from from sources such as water or fossil fuels. Today this requires electricity, so it isn't yet the clean cure-all we seek. Still, it can't be overlooked, because it has the potential to be just that.[220]

President George W. Bush declared his dedication to pursuing a hydrogen future in his 2003 State of the Union speech, unveiling his Hydrogen Fuel Initiative. In doing so, he identified the most vexing problems preventing the use of hydrogen on a massive scale, saying:[221]

> *"To be economically competitive with the recent fossil fuel economy, the cost of [hydrogen] fuel cells must be lowered by a factor of ten or more and the cost of producing hydrogen must be lowered by a factor of four ... Performance and reliability of hydrogen technology for transportation and other uses must be improved dramatically ..."*

Lowering the cost of producing hydrogen is an obvious objective, but first, it's important to highlight the big problems that President Bush didn't mention, as well as the opportunity President Bush didn't recognize. The current methods of producing hydrogen require the use of a great deal of energy and, when this energy is produced from fossil fuel, they also result in the release of greenhouse gases into the air. Ultimately hydrogen can be a viable replacement for all the forms of energy we now use. But that will require discovering a way to separate it out of water with the heat of the sun and some benign chemical process. Then we can produce it in its pure form at mass scale and a competitive cost that does not contribute to the buildup of greenhouse gases. We need renewable-produced hydrogen.

The most common method of obtaining pure hydrogen today is using steam to heat fossil fuels—primarily oil and natural gas—until the molecules in which the carbon is bound up break down, allowing the hydrogen molecules to be extracted. This process is like a dog chasing its tail. Increasing this type of production will only burn more fossil fuel and add to the emission not only of carbon, but of methane and other particulates or to the creating more toxic nuclear waste. This is why policy makers seriously concerned about climate change, generally have a rather skeptical, and even negative, attitude toward initiatives to use hydrogen as currently obtained.[222]

There are two ways that hydrogen can be part of our clean energy future. First, hydrogen is a good way to store solar or wind power. This is done by using the solar- or wind-generated electricity in a process called electrolysis to split the hydrogen in water, which can then be used when needed in a peaking power plant, or as a substitute for natural gas or oil heating. And, of course, in a fuel-cell car. But storing the solar and wind as hydrogen doesn't add to our energy supply.[223]

What would really make hydrogen the answer to our clean-energy future is success with research into pollution-free and energy-free ways to extract hydrogen from water. Such a research project is being conducted at the California Institute of Technology and elsewhere. The researchers are attempting to find a method that relies on the direct heat of the sun to separate the hydrogen from water. The effort is as of yet on a small scale and is showing progress, but it hasn't received much attention. If a breakthrough is achieved, we could find ourselves in the "Age of Plentiful Clean Energy" faster than we would have thought possible.[224]

The prospects of such a breakthrough are possible, and given how momentous that would be, we feel compelled to promote the "easy hydrogen" option. And we believe it should be pursued with the same fervency behind the Manhattan Project and the race to the moon. Both of those achievements seemed near impossible and were achieved in a matter of only a few years.

More Renewable Hydrogen Now

Even as research into such a breakthrough proceeds, hydrogen extracted by solar and wind powered electrolysis can start replacing oil and natural gas. In order for that to happen, we must start building renewable hydrogen production centers. One idea for this is to build production facilities right at the solar plants and wind farms. Indeed, some of the large projects may face less opposition if their overhead electric transmission lines to move electricity across deserts and beautiful mountains were replaced by underground pipelines to transport the hydrogen gas. But no major increase in production will happen if we don't begin developing a market for renewable hydrogen that includes the necessary infrastructure of hydrogen "filling stations," and the hydrogen-burning cars to use them.

The technology is available to start building the "hydrogen highway" that was proposed by then California Governor Arnold Schwarzenegger back in 2005. He signed an executive order calling for the building of two hundred hydrogen filling stations in the state. Although there has been some renewed interest in California, they weren't built, and the interest expressed in New York State has suffered a similar fate.[225] Hydrogen faces the same chicken and egg problem that has slowed adoption of electric cars—until more cars are on the road, investing in building the filling-station infrastructure to fuel them isn't cost-effective, but mass adoption of the cars isn't feasible without the fueling infrastructure. This is a large part of why commercially sold hydrogen fueled cars are not yet a reality, even though the technology is adequate to produce them. There has recently been some attempt to increase the number of filling stations, yet inertia remains.[226]

For electric cars, the utilities have at least begun to build charging stations at your home and around town and the cars are commercial. Yet with the highly desirable hydrogen car, which can travel as far as a gasoline car on a tank of fuel, we still see only demonstration models. Some more viable options have been tested, though. In 2003, Ford unveiled a Model U hydrogen hybrid SUV, a concept car that had an internal combustion engine that ran on hydrogen and an electric motor, with a three-hundred-mile range before refueling was required. It would have been a worthy successor to the life-changing Model T, and the Ford engineers said production of a hundred thousand vehicles per year would have brought the sales price down to about that of gasoline cars. But though the engineering of the car was a tremendous achievement, the company has not pushed forward with any plans to mass-produce it. Unlike the Model T, the Model U has been forgotten.[227]

There was some local government interest in hydrogen a decade ago. The city of Santa Monica, in Southern California, installed a hydrogen fueling station in 2006 and bought five of the Toyota hybrids. The South Coast Air Quality District in California retrofitted a few cars to run on hydrogen. But the Hydrogen Initiative of President Bush appeared to have died aborning.

Although the Obama administration has not advanced "hydrogen," only recently giving some belated support to hydrogen vehicles, the automobile industry, nevertheless, has continued efforts to develop an affordable hydrogen car.[228] In 2014, the California state government granted some $50 million to build a few hydrogen filling stations, mainly in the south of the state.[229] In the spring 2005, Toyota leased thirty hydrogen-fuel cars to selected customers in southern California, and in 2015 Toyota is set to introduce its first hydrogen fuel-cell car for purchase in the U.S., so there is life in the hydrogen option after all.[230]

The Hydrogen Fuel-Cell Car

The auto industry understands that the hydrogen fuel cell is an excellent piece of technology. A by-product of the space program, it's an ingenious device that chemically converts hydrogen into electricity.

A fuel cell operates in the reverse of the process of electrolysis; while electrolysis uses electricity to split water and create hydrogen molecules, the fuel cell breaks down hydrogen molecules to create electricity and water, at least twice as efficiently as an internal combustion engine. The electrons are stripped from the hydrogen molecules and run through a circuit, which generates electricity for running the car. And the only by-products are water vapor and some heat. Like a solar panel, the fuel cell produces electricity without requiring any moving parts.[231]

A distinct advantage of hydrogen fuel-cell-powered cars over electric cars of today is that the fuel cell overcomes the range limitation of electric batteries. The fuel-cell cars can easily go over three hundred miles without refueling, while today's all-electric cars are reliable for only one hundred to two hundred miles. The exception of course is the Tesla which, with the largest battery array, goes nearly three hundred miles between charges. A few other manufacturers, such as Nissan, may come out with longer ranges, near two hundred and fifty miles.[232] Despite the EV impending gains in distance, fuel cells find it easier to reach a large mileage range and fuel cells are also much smaller and lighter than electric batteries, adding to fuel efficiency.

But as of now, fuel cells are not an option for the general public. They are not for sale or lease except in tiny numbers by Toyota and the "hydrogen highway" hasn't been built even in California. The American public doesn't know that renewable hydrogen is an option that is begging to be made a commercial reality.

The Safety Issue

One of the main concerns raised about hydrogen has been that it might explode. The safety question must be taken seriously, and it has been by the engineers working on the technology. The tanks to hold hydrogen gas under pressure have been designed with safety as their primary objective. They have withstood extensive testing. As with gasoline tanks, hydrogen fuel tanks are designed to make an explosion nearly impossible, and they've been certified as safe by government authorities.[233]

Some people recall the 1937 fireball disaster of the hydrogen-filled Zeppelin Hindenburg and think hydrogen in our cars and in storage tanks at filling stations poses the risk of such explosion. But the fact is that the fire that consumed the Hindenburg wasn't due to the hydrogen inside the dirigible, but to the high flammability of the outer material. If hydrogen is released, because it is lighter than air, it simply evaporates into the atmosphere.

In fact, hydrogen is safer than gasoline. Gasoline is a highly flammable liquid that coats surfaces and causes fires following collisions which kill hundreds of people in the country every year. Hydrogen is actually less combustible than gasoline and disperses into the air. Nothing is absolutely safe and no one can say that hydrogen could never explode. But if safety is the issue, hydrogen cars will be safer than today's cars.[234]

Solving the Chicken and Egg Problem

The key to creating the major market for hydrogen is to overcome the notion that we must build enough fueling infrastructure to support mass adoption before we get going. Critics have argued that this infrastructure will cost many billions of dollars. That assertion must be more thoroughly investigated, and the calculations must also factor in the value to society, national security, public health, and investment opportunities. In addition, this complaint overlooks that we've made an enormous national investment in our motor vehicle infrastructure before and reaped great rewards.

The federal interstate highway was a tremendously expensive project advanced by President Eisenhower with the help of Senator Al Gore Senior. Building the network of two hundred thousand petroleum fueling stations we enjoy today was also a major undertaking. What if Henry Ford had let the lack of highways and filling stations deter him? When he built the Model T there were few if any paved roads, let alone highways, and no gas stations at all. You could purchase gasoline only at a drugstore.[235]

We don't have to build the network of stations all in one fell swoop. Both cars and filling stations should be built on a coordinated schedule, starting with a few stations and a few vehicles in a few locales and growing from there. Toyota, Honda, and the State of California are showing an example of how to start. Now we need to accelerate the pace.

We could begin by purchasing hydrogen-fueled cars and trucks for city fleets, ranging from police cars to taxis and garbage trucks. Rental car companies could show some leadership. Fueling stations could be built and operated by the cities, and for the rental car companies, the fueling facilities could be located in their lots. There is every reason to believe that by 2020, a critical mass of hydrogen infrastructure for such fleet operations can be built. In the plan we present later, to spur wider adoption by consumers, we recommend a sizeable tax credit for both car owners and manufacturers and comparable incentives for cities as well.

Hydrogen Beyond Cars

Other ways in which we can use hydrogen fuel are for powering aircraft and ships. A number of studies have shown that a fleet of airplanes fueled by hydrogen could be part of our future. The European consortium CRYOPLANE project concludes that, "Hydrogen could be a suitable alternative fuel for future aviation; based on renewable energy sources it offers the chance to continue the long-term growth of aviation without damaging the atmosphere. Importantly no critical barriers to implementation were identified in the study. Further research is needed, but implementation could take place within fifteen to twenty years."[236]

Hydrogen is appealing for fueling planes because it is much lighter fuel than petroleum. Research has shown that hydrogen fuel would enable jets to fly nonstop around the world. The planes would have to be wider in order to store the amount of fuel needed, but studies conducted twenty-five years ago showed this is feasible.[237]

For ships, hybrid hydrogen and battery systems are being developed by the research arm of the shipping industry. One of these systems uses liquid hydrogen as the main fuel, and then a combination of a hydrogen fuel cell and a battery system, to conserve space on the ship.[238]

Hydrogen fuel cells need not be confined to motor vehicles. They can be used to generate electricity for both industries and homes. Hydrogen could also be used for home heating and cooking.[239]

As previously mentioned, hydrogen can also work into the all-renewable system as storage. Excess solar and wind generated electricity could be used to separate hydrogen out of water and that hydrogen could be stored and then burned to generate electricity for peak usage.[240]

Further down the road, it may well be that a decentralized approach of making hydrogen in relatively small quantities wherever there are sources of water and renewable-generated electricity will be best. Imagine a solar-panel-powered electrolysis machine and hydrogen storage tank and dispenser in your backyard. You would be able to heat your home and power all of your appliances and could fill up your car whenever needed, never needing to go to a service station again. Honda has done extensive research for the decentralized hydrogen approach.

The Hydrogen House Project participates in showcasing a home that runs on a solar-hydrogen system. Having remained off-grid for eight years, the house "provides a complete energy picture with energy generation, on-site storage and reconversion, all from renewable resources. All the appliances and vehicles down to the lawnmower have been converted to operate using renewable hydrogen gas." This is a vision of what the future could be.[241]

But developing hydrogen technology won't happen until the public and public officials realize that renewable hydrogen—like renewable electricity—needs to be part of our energy future.

To set the transition in motion in earnest, we're going to need to several things:

1. Enacting the incentives and funding to make the hydrogen fuel-cell car a commercial product.

2. Begin building the hydrogen highway in earnest.

3. Build several large complexes where solar and wind power are converted to hydrogen and then pipe the hydrogen to markets.

4. Make the research commitment to split hydrogen from water, without using electricity, equal to that of the priority we gave to the Manhattan project in the 1940's.

CHAPTER 11

Electric Cars and Heat Pumps for Every Home

Heat pumps are an often-overlooked and essential resource for substituting renewable energy for natural gas and oil and heating buildings. They can be used not only for heating and cooling our buildings and heating water, they are important players in addressing the thirty-five-year deadline for stopping the release of greenhouse gas emissions from our buildings and industries.

How do they work? Unlike natural gas furnaces, which burn fuel, and electric generators, which heat up coils to generate heat, heat pumps extract heat from a low-temperature renewable source such as, the ground, water, air or from industrial or domestic waste, and transfer it to a building or industrial application.[242] Run on electricity, they achieve great energy efficiency and are considered a renewable technology because of their ability to harness the solar energy stored in the earth, air, and water.[243] The result is a renewable system that provides cost savings, energy efficiency, and a way to heat and cool without the use of fossil fuels.

A good way to understand the technology is to look at refrigerators. Heat pumps are essentially refrigerators and air conditioners running in reverse to provide heat. In fact, a great positive feature about them is that they're a single unit that can heat, cool, and provide dehumidification. They are also able to provide heating and cooling for varied industrial processes. Another benefit is that heat produced when cooling can be used for other needs.[244]

Our Buildings Are Huge Energy Wasters

Moving to a twenty-first century infrastructure in the building sector is as important as in the transportation and electrical generation sectors. Research has shown that, "buildings consume approximately 40% of the United States' and world's primary energy and are responsible for 40% of greenhouse gas emissions."[245] Our buildings consume more energy than the roughly 30% we use to run our entire transportation sector.[246]

The majority of our heating is done with natural gas, propane, or oil. There are also some direct electric heaters for heating. They are all highly inefficient when compared to heat pumps. In 2009, air conditioning, combined with the heating of our residential homes and water, accounted for 65.4% of all of the energy we use in our homes.[247]

The natural gas industry argues that their highly efficient furnaces are more efficient than electric heat pumps if the electricity is made with fossil fuels. Studies find that this is simply not true for today's energy supply.[248] However, this no longer really matters. That is yesterday's debate. For tomorrow, electricity will be made from renewable energy and heat pumps will emit zero greenhouse gases, which is impossible for the efficient natural gas furnaces. The natural gas furnace will forever emit huge quantities of carbon and be fed by a supply chain that leaks methane.

Today's issues are clearly whether heat pumps can heat buildings in cold climates and whether the cost of their heat is competitive with natural gas and propane in our furnaces. The answer to both questions is yes.

New and Improved Pumps

Why aren't more homes and businesses installing heat pumps? The fundamental answer is that the electric power industry has not promoted them. They have sat by and let gas and oil capture the market. In the past, the prevailing view was that heat pumps were too expensive, unable to perform in colder environments, and too complicated to install. However, the technology has evolved to a point where these concerns are no longer true.

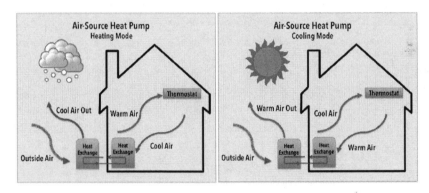

FIGURE 8: Air-Source Heat Pumps. Copyright: All-Electric America.

The most common types of heat pumps used for both residential and commercial purposes are air-source and ground-source heat pumps (often referred to as geothermal heat pumps) seen in Figure 8 and Figure 9. Both types can be used not only in warm climates, but also in climates as far north as Canada, Scandinavia, and Alaska.[249] They can replace electrical baseboards, oil, natural gas, kerosene, and propane for heating. They can also provide cooling in the residential, commercial, and much of our industrial sectors throughout the United States and the world. Today's heat pumps provide heating and cooling at prices comparable and even less than today's other heating systems.[250]

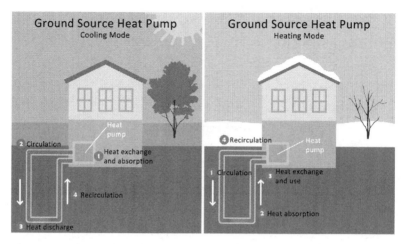

FIGURE 9: Ground-Source Heat Pumps. Source: http://www2.epa.gov/rhc/geothermal-heating-and-cooling-technologies.

Ground-source heat pumps are the most efficient and are the least expensive heating and cooling units to run over the lifetime of the unit. Though they've been proven to excel in cold climates and have low operating costs, they are more complicated and expensive to install.[251] While they have a payback time of five to ten years,[252] the upfront costs can be prohibitive. Upfront financing by utilities and incentives are needed. With proper marketing, financing, and incentives, though, they will make the ideal heating option.

While previous generations of air-source heat pumps have been easy to install, below 25 °F (-4°C) they became ineffective and required extensive backup heating. Thus in the coldest climates they produced limited savings.[253] However, today there are new options. The latest generation of cold climate air-source heat pumps can replace gas-fired furnaces throughout North America, including Canada and Alaska.[254] Cold climate air-source heat pumps use far less energy than a gas furnace and are equivalent or less expensive to run than gas furnaces in most locations throughout the U.S. The ease of installation and upfront costs are comparable in price to natural gas systems, even without full commercialization.[255]

Another plus is that heat pumps provide a predictable rate of electricity consumption for heating and cooling. Electricity rates are highly regulated and are not as subject to the whims of the ever-fluctuating cost of natural gas and other fossil fuels. Solar units can be coupled with these systems to add additional benefits by reducing a home's electric demand from the grid. Moreover, heat-pump units do not require a lot of equipment and last twenty or more years, with ground-source coils lasting roughly fifty years.[256]

It would be in our governments' best interest to provide tax incentives and financing to help consumers with the initial cost of heat-pump installation. It would also be of even greater value for our utilities to promote and even finance the installation of pumps. Some utilities and state and federal organizations have already recognized their value. Tennessee Valley Authority (TVA), Oak Ridge National Laboratory, and the state of Maine are all leaders in promoting and helping the customer acquire this technology. Programs and incentives to encourage their use are described in more detail in the next section.

Initiatives by Utilities and Government Agencies

The Tennessee Valley Authority has been a leader in this effort for over sixty years. Partnering with local public power companies, TVA created the EnergyRight Solutions Heat Pump Program, which allowed customers to make low monthly payments, added to their electric bills, and to pay over the course of ten years. TVA also provides access to a Quality Contractor Network (QCN) for installing the pumps.[257]

In recent years, TVA has partnered with Oak Ridge National Laboratory (ORNL) in developing new advanced heat pumps paired with renewable energy. Their projects include the invention of a frostless heat pump integrated with a photovoltaic system seen in Figure 10, which provides greater thermal comfort, and the promotion of heat-pump water-heater technology seen in Figure 9, which is 50% more energy efficient than traditional water heaters.[258]

FIGURE 10: "Projects involving TVA and ORNL include (left to right) integrating photovoltaics in roofs, purchasing of TVA's "green power" by ORNL, and promoting frostless heat pumps." Source: Oak Ridge National Laboratory, http://web.ornl.gov/info/ornlreview/v38_1_05/article13.shtml.

FIGURE 11: "Projects involving TVA and ORNL include heat pump water heaters." Source: Oak Ridge National Laboratory, http://web.ornl.gov/info/ornlreview/v38_1_05/article13.shtml.

The state of Maine is also a leader in promoting heat pumps through its Efficiency Maine program. Maine residents today can lock in the price of heating and cooling with a geothermal unit for twenty years with no upfront cost. A typical ground-source unit costs roughly $40,000, however, the consumer is able to access Federal tax breaks at about $12,000 (30% installation cost) and the efficiency Maine incentive at $5,000. The $23,000 balance of the geothermal system is then financed by Efficiency Maine with a PowerSaver loan for $182 per month, ($2,184) per year for a fifteen-year term at 4.99% APR fixed. This all adds up to a 9.4-year payback time with savings of $2,450 a year and $49,000 savings over twenty years.[259]

Other state and utility initiatives around the country include:

- Vermont Green Mountain Power (GMP) rents cold-climate heat pumps to customers. Its cold-climate heat-pump services allow customers to save up to 50% on energy costs with GMP's rental program for as low as $39.99 a month. GMP provides quick and easy installation with no upfront cost. Their complete energy makeover program is discussed in detail in Chapter 5. [260]

- In January, 2015, New York Councilman Costa Constantinides introduced a bill to establish a series of guidelines for commercial and residential use of geothermal energy, and to build a directory of city buildings that might be best suited for geothermal (bore-holes).... Since 2012, the New York City Housing Authority installed geothermal projects for 225 units of public housing."[261]

- In late 2014, State representative Aric Nesbitt introduced a bill to the Michigan House that proposed including geothermal heat pumps in an expanded definition of "Renewable Energy," making them available for utility Renewable Energy Credits.

- In 2012, New Hampshire's legislature passed and Governor John Lynch signed a law allowing "thermal renewable energy sources to be used by electric utilities to meet energy purchase requirements."[262]

Not only are these initiatives to be applauded, they should be replicated by every state in the country, and they should be supported by the federal government as well. The greenhouse gas reductions in energy savings we can achieve are simply too important for all families and all commercial building owners to miss out on.

Electric Cars

The electric car is not a new idea. There were electric taxicabs in New York city in the 1890s. Range was not a problem because the batteries were exchanged for fresh ones at the electric "filling station." Electrics were the predominant motor vehicle until Henry Ford came along with the Model T and left them in their dust.

Various electric car models have been tried out for decades. In the 1970s General Motors and Ford were promising commercial electric cars. Dave, as Chairman of TVA, received a letter from the then president of General Motors in 1978 promising to have an electric car available in every GM showroom by 1984. He didn't keep his promise. However, GM did produce the EV-1 in the early 1990s, only to recall and smash them to death a few years later. The documentary "Who Killed the Electric Car" documents this hard-to-believe story.

Today GM, Nissan, and most car manufacturers have some version of an all-electric or a hybrid-electric car on the market. Sales are disappointing. The reason cars from these companies are not selling is that they cost too much and don't go far enough without needing to have the batteries recharged. Charging stations have not been built so that a charger is available close by wherever you are.

Then Tesla arrived on the scene, a new car company that sells only electric cars. It is pursuing a "trickle-down theory" starting with very expensive cars that do have a longer range and promising that in a few short years the price will come down to be competitive and in the price range that most of the general public pays for new cars.

A major obstacle to massive sales of electric cars is the dealers themselves whose profits hinge on servicing the cars with internal combustion engines that need lots of maintenance. They, more so than most consumers, understand that electric cars have very few moving parts, except the wheels, and are going to be pretty much maintenance-free and have a longer life than existing gasoline cars. That is one reason why Tesla is selling their cars directly and not through distributors.

The State of California, in the early 90s, understood that the automobile companies left alone had little incentive to develop a new line of cars that would compete with the gasoline cars they were already selling. So, Republican governor George Deukmejian signed into law the requirement that automobile companies sell an increasing percentage of zero-emission (electric or hydrogen fuel cell) cars each year. Unfortunately that law ceased to be implemented when the cars did not become available.

Today, the electrics are available, but until the price is reduced, the range for all of the cars is increased to 200-300 miles, and the promotion gains more strength, electric cars will not be on the road in the numbers required by the climate crisis. Until these goals are accomplished, the health hazards from burning gasoline at lung level in the city streets will remain.

The march of progress in overcoming price and range problems is impressive, but what is lacking are the combination of requirements, incentives, and promotion needed to see massive deployment of electric cars.

The Chinese government decreed that there will be five million electric cars on the road in China by 2020.[263] Surely the United States of America can match that number at the very least, but it will not happen unless we go forward with the following actions:

1. Require that an increasing percentage of all cars sold in America be zero-emission vehicles—electrics, hydrogen fuel cell, or any other combination of technologies that meet that requirement.

2. For at least an initial five-year period, provide a 20% tax credit for zero-emission vehicles to assure the price is in fact affordable for massive sales.

3. Inspire the electric power industry to start promoting the sale of electric vehicles by owning the batteries, installing charging stations, and offering bargain off-peak rates.

Switching to electric cars is an important way for us to achieve energy efficiency. The general public has not been told that over the life of an electric car it is cheaper than a comparable gasoline car for three very important reasons:

1. The fuel will cost less than half as much as gasoline. Not only will it use significantly less energy, but we can expect to pay roughly one dollar "per gallon" in electricity.[264]

2. Maintenance costs will be a small fraction of repairing internal combustion cars.

3. The resale value of electrics will be higher; especially if the utility owns the batteries because nothing else in the car will wear out.

The electric car offers a tremendous growth opportunity for the electric power industry. Here is a chance for the utility to promote a technology that will grow their business, clean the air locally, increase energy efficiency, combat climate change, and save consumers money. Hopefully, they will embrace that challenge and make it happen.

This is a product that will actually be more fun to drive because it can give you quiet jackrabbit start-up and acceleration as well as awesome handling—pleasures usually associated only with the most expensive cars. Imagine New York City, Chicago, or even Los Angeles free of the smog and noise now caused by today's vehicles. Even the electric power industry ought to be able to market that product.

CHAPTER 12

The Silent Threat of Too Much Stuff: Make It Built to Last

The climate challenge should inspire a fundamental reassessment of America's addiction to increasing our material affluence. Our economy depends far too much on Americans buying more and more stuff they really don't need. Let's face it, the key features of present-day middle-class American life are unrestricted use of large family cars, the comfortable enjoyment of an oversized suburban home situated far from mass transit and shopping districts, and ever-increasing consumption of material goods. We have set a bad example that China, India, and the developing nations are imitating.

America's preoccupation with material goods goes to the heart of the climate problem. Making goods to last and being less wasteful will make it easier to achieve the transition to our all-renewable world. So America's leadership in combating the climate problem should include a shift in mindset away from our materialism to an appreciation that less can be more and that we'll all be better off if we build stuff to last. This can be viewed as another form of energy efficiency. We will use less energy, for example, by making durable goods and repairing them when they break instead of just throwing them away and buying new ones.

Breaking our consumption addiction require a change in our values. The good news is that politicians, social theorists, and ordinary citizens throughout the developed world are questioning the material growth ethic, seeking a more balanced concept of both national development and individual fulfillment. The many Americans who have embraced a "voluntary simplicity" lifestyle, as advocated by author Duane Elgin, have found that happiness does not depend on more and more material goods.

Many times in our history we have made significant adjustments to our lifestyles. Americans have voluntarily reduced the country's birth rate drastically, for example. When we faced the oil crisis of the 1970s, we learned to carpool, to insulate our homes, and to turn down our thermostats. We've made great strides in recycling and using reusable shopping bags. We can do this. And as was true in each of these other cases, we will become stronger for it.

Of course, many Americans have yet to attain a decent standard of living, much less a life of affluence. For them, economic growth is an urgent necessity. But for those of us who have achieved a high standard of living, learning to reign in our consumption and focus on quality of goods rather than quantity should be the goal.

Fessing Up to Economic Fallacies

A root fallacy in current economic analysis of the strength of our economy is in how we measure productivity. A key metric of success is the number of automobiles or shoes or light bulbs produced per unit of investment in capital and labor. Yet, if we think about it, the number of light bulbs a worker can produce in a day and the total manufacturing cost of a bulb are not the decisive tests at all. What is more important is how long the bulb will last and its cost per hour of use for the consumer. We should measure productivity instead by the unit cost of service to the consumer.

Another fundamental error in the prevailing economic measures is that many of the production costs of industry are not accounted for. Those costs should include damage to the environment and to human health, as well as the costs of disposing of poorly built automobiles, toasters, or tricycles after they fall apart. We do not even calculate the cost to dispose of one-time use items such as paper napkins, disposable diapers, fast-food wrappers, or the plastic packaging used ubiquitously at drugstores, mini-marts, and box stores.

Our economic system has performed quite well in achieving our current standard of living. But it is time to put in place incentives that would encourage industry to make products that last longer, are repairable, and can be recycled and reused. To be sure, often people buy new things not because the old ones have worn out, but because the fashion changes or a new model is available. That's not going to change. But much of what America buys isn't subject to shifting fashion or regular product updates. Light bulbs that last longer have caught on well, and so would any products built for better longevity. Many items are also discarded because there is no easy way to repair them.

A realistic and desirable objective for the U.S. would be a full-employment, knowledge-intensive, food-and-service-oriented economy that would be fueled at a fairly stable level of energy consumption using renewables. This economy would still afford a high degree of material affluence and would still consume huge quantities of energy, but it would do much less damage to the planet.

Achieving greater durability in the products would affect the demand for energy in a fundamental way. If energy-intensive industrial products lasted twice as long as they do now, annual energy requirements to make them could be cut in half.

Changes in the manufacturing process to achieve such savings need not seriously disrupt the economy. Economic efficiency would still be the test, but it would be measured over the life of the product and on the basis of the full cost to society. If all manufactured goods were accurately labeled to reveal their expected (or better still, guaranteed) life-span and their cost per unit of useful service, we might see companies competing with more focus on winning a public reputation for quality craftsmanship and longevity. A tax incentive that increased according to the length of time a product was guaranteed to last might cause a major shift in production methods. Regulations could be put in place to require that products are built so that the materials they contain could be 100% reused. Equitable freight rates for transporting recycled materials could complete the package.

The apparel company Patagonia has shown that transforming the manufacturing process with these goals in mind can in fact work, and work quite well. So serious is the company about its commitment to conservation that it ran an ad campaign in the New York Times seen in Figure 12 that encouraged consumers to buy less of the company's merchandise on Black Friday. The ad also described the harm done by energy-intensive industries worldwide. The company's Common Threads Initiative is aimed at helping consumers "reimagine a world where we take only what nature can replace," promoting the repair, reuse, and recycling of their products in order to cut down on energy consumption.

A movement promoting a new emphasis on quality should have strong appeal to consumers, offering the promise of a better return on our investments in goods and better service in maintaining and repairing them. On top of this the talents of workers in the mass-production plants, whose days are devoted to a mindless drudgery, could be much better utilized, and their work lives vastly improved, if production processes involve more quality assurance.

FIGURE 12: Patagonia Common Threads Initiative

Better Jobs, Not Fewer Jobs

A common claim made by those arguing against energy conservation measures for industry is that they will reduce employment. Here it is important to distinguish between planned conservation and a sudden unexpected shortage. Obviously if industry is suddenly deprived of energy it has counted on, production slows down and jobs are lost. But a gradual shift to better-insulated homes, more efficient cars, and less energy-intensive industrial production is no threat to employment.

America has witnessed an increasingly postindustrial economy that has lead to fewer people working in factories. But other higher-quality jobs can fill the void. An emphasis on reuse and repair would create a need for skilled services in repairing and servicing of products. The net effect would be more green-collar jobs to replace the blue-collar jobs we have already lost.

Life Will Be Better, Not Spartan

Some Americans argue that there is no role for government in industry, and that regulations hamstring companies, increase costs, and stifle innovation. They also argue that the government has no place in regulating our personal lives; that telling us, for example, to drive more energy-efficient cars and to use mass transit is coercive. None of that is necessarily true; it all depends on the nature of the regulations. There is nothing coercive about a society in which buildings do not leak and people get to work by rapid transit rather than by driving single-passenger cars; where short trips are taken by high-speed rail rather than by driving or flying; and in which the products of industry are built to be durable and recyclable.

On the contrary, instituting regulations that further these goals is entirely in keeping with our county's founding ideal of promoting the pursuit of happiness.

Part IV
Let's Make It Happen

CHAPTER 13

From Metropolis to Ecopolis

The ancient Roman poet Horace proffered the sage advice "carpe diem," meaning "seize the day," expressing the timeless wisdom that it's best to take full advantage of the opportunities of the present moment rather than putting things off for another day. Understanding which actions to take, however, can be difficult, especially with such a complex and interconnected set of parts as those that make up our energy supply. So in this chapter, we will use the city of Los Angeles as an example of how a concerted, well-calibrated transformation process to an entirely renewable power system can be executed in a major metropolitan area, even one that is perhaps the most car-crazed of all major world cities. If Los Angeles can kick its oil habit, then it's entirely reasonable to believe that New York, Dallas, Shanghai, Mexico City, Jakarta, and all the other metropolises around the globe can as well, not to mention smaller cities and towns, which pose less daunting challenges.

We've chosen LA not only because it's so notoriously car-crazed, but because Dave knows the city's power system well, having worked as general manager of the LA Department of Water and Power (LADWP) from 1997 to 2001, and was deputy mayor in 2010.

Also, because the city's leaders have already expressed a desire to make the city greener, LA has a green vision, which is more a dream than a reality to date. However, such vision must be translated into action on the part of civic leaders along with public demand, without which even the best of plans will make little difference. We hope that readers will think of their own cities and towns, and states as a whole, as they read this scenario, and consider point by point how similar changes could be enacted where they live.

Greening the Electrical Grid

The basic plan should be to transform the city's electrical generation capacity to solar, wind, geothermal sources, and the existing hydro-power capacity all backed up by energy storage. Renewable energy is too precious to waste so massive investments in using it efficiently must be made so the customers can have lower energy bills than they pay today. On a parallel track, LA needs to become the home of electric and hydrogen fuel-cell cars. Solar hot water and electric heat pumps can replace natural gas, and fast electric trains can replace most jet-fueled air travel within California. Fortunately the city owns its electric utility, the LADWP, which gives the city's leadership a strong hand in enacting these changes. The city also has vast resources available.

A Solar City

LA is, of course, lucky enough to be bathed in sun all year round, and solar generation is the primary source that should fuel its transformation. The extraordinary opportunity for massively increasing its production of solar power stares you in the face as you fly into the city. Sprawling out over an area of 502.7 square miles, the city has an extraordinary acreage of warehouse rooftops alone on which to install panels.[265] With the roofs of commercial buildings and houses added to that steadily over time, the city could become a massive solar producer. Imagine flying into LAX and seeing a vast array of solar collectors glinting on the roofs of virtually every building and the many sun-drenched parking lots.

Watching the Electric Meter Spin Backwards

The city has begun an aggressive policy of encouraging rooftop solar PV as it takes steps in implementing the mandate of the 33% Renewable Portfolio Standard by 2020 and 50% by 2030. The LADWP has expanded a demonstration feed-in tariff program, where the utility pays for customer-generated solar, which now expects to generate 100 MW of solar power by 2016 and 150 MW by 2020. This program is targeted to larger customers who install permitted solar arrays which will sell their power to the grid at fixed prices.[266]

The city could greatly accelerate the installation of roof panels by asking its customers to "lend their roofs," installing city-owned solar systems for integrating the electricity generated into its grid. Because the city-owned utility has nonprofit status, low-cost municipal financing is available to fund the program. But the LADWP has the authority to institute such a program on its own. If the green vision is to become a reality, city leaders need to do just that.

A LADWP program to market rooftops and microgrid solar systems would not only result in lower costs, but assure that lower-income citizens would enjoy the benefits of solar power. Without such a program, those who need lower-cost electricity will be stuck with higher-cost electricity than those who can afford solar systems on their own.

Imagine the smiles on the faces of consumers as they watch their electric meters run backward when surplus energy from their solar panels is automatically sold back to the LADWP. Public enthusiasm would boom. To make this a reality, LADWP must value the power generated at the retail rate for the rest of the electricity it generates. The utility has already taken steps in this direction with the expanded feed-in tariff program. But that program could expand, particularly to smaller customers. And it counts on tax credits that very well may expire in 2016.[267] The truly low-cost option is municipal bond financing by the nonprofit LADWP.

Plenty of Big Solar Potential

LA has already identified sites where solar fields could be installed on land it owns that could generate thousands of megawatts of power. One especially big resource is an undeveloped expanse of land the size of the city itself two hundred miles to its north in the nearby Owens Valley. It was the water from Owens Lake that enabled LA to grow from a village to a big city. The sun beats hot year-round there, and there are vast stretches of more than a hundred square miles of empty space in the former lake bed and near the already existing transmission corridor.[268]

Building solar plants in the Owens Valley would require consultation with the local communities, bringing them in as partners in site selection and assuring an equitable distribution of the benefits. Transmitting the electricity to the city would also require expansion of the existing transmission corridors. In 2014, LA finally made peace with the residents of the Owens Valley over LA's use from the Valley. The dust from the resulting dried lakebed will be controlled. Now is the time to enrich the local economy by harvesting the solar gold mine that keeps on shining every long day.

The solar farms that can be built near existing transmission lines combined with rooftop solar could meet ALL of LA's need for electricity if combined with large investments in storage. Thus far LA has relied on the fact that it owns a sizeable pump-storage plant. But as it expands beyond the 20% renewable power of today, storage must be built to make the renewables a larger and larger source.

But solar doesn't have to do it alone. The city has many additional good resources for renewables.

The Whistling Winds

LA is ringed to the north by the Tehachapi Mountains, which offer abundant wind, as does the southern region as a whole. The California Energy Commission estimates a total wind potential of at least 4.5 million kW in Southern California.[269] And LA isn't confined to just these relatively close sources. The city owns extensive electrical transmission lines to connect it to the enormous wind resources in Wyoming and adjacent states. The yet-to-be-built potential is over 500,000 MW, which dwarfs the entire LA electric capacity in 2015 of 7,200 MW.[270] One wind farm alone in Wyoming able to produce 4,000 MW is under active development for sale to California and has been offered to LA.[271]

None can question that abundant wind and solar power are available to Los Angeles, which combined with storage, can substitute for coal-fired plants. The greatest challenge facing the elected leaders of LA is whether they replace the coal plants with renewables and storage or implement the current plans to build natural gas plants—plants that have been documented to be as dangerous to our climate as coal.[272]

Garbage Transformed into Treasure

Possibly the biggest win-win for a city in power production would be taking the mountains of municipal waste that no one wants to live near and converting them into electricity, or perhaps down the road, even hydrogen. The technology for doing so is available, and plants could be built economically. This technology is a far cry from the highly polluting waste-to-energy plants of the 1970s. New waste plants like these are in use in Germany and much of Northern Europe. They would work just as well in LA.[273]

A process has been developed for heating organic waste in an anaerobic environment, meaning one without any oxygen, and converting it to hydrogen, which is called pyrolysis. Building a number of facilities to do both would be economical because they could operate around the clock and not require any storage. The economics make particularly good sense when you factor in the huge cost to the city of managing its waste disposal. LA is presently shipping much of its waste to a landfill in a neighboring county. That is not a permanent solution. Turning this waste into fuel is a better solution. Some small-scale efforts are under way but modern-day waste to energy isn't yet a commercial reality. The city could give a decisive boost to development with a commitment to building one plant of 50 MW per year or more. At that pace, in fifteen years' time (allowing five years for start-up) the city could be generating 500 MW of electricity from municipal waste, and within twenty years, 750 MW, enough power to keep the lights on for 450,000 homes. Municipal waste could supply 5-10% of total electricity or be a major source of hydrogen for fuel-cell vehicles.

The best estimates that Dave found in his book, "Winning our Energy Independence," in 2007 was the total power the city could ultimately generate from all of these sources were: 5,000 MW from solar, 13,000 MW from wind, and 750 MW from municipal waste, for a total of 18,750 MW. LADWP could thus triple its output of electricity to replace oil and gas in LA with a greenhouse-gas-free energy supply.

Transforming Usage

Of course creating the additional renewable electricity needed is only one half of the equation. A steady switch to electricity from cars that run on gas and homes and water heated by natural gas is the other half.

Distant goals will only be achieved if the transition takes place one year at time, every year. The power plants that are powered by coal, natural gas, gasoline, and nuclear power can be phased out about 3% a year at the same rate of buildup of renewables. And additional storage must be added each year as well. The transition can be seamless. This would in no way jeopardize the reliability of the power supply or require the city to buy power to fill gaps. Indeed that transition has already begun as LA and the rest of California went up to 20% renewables by 2010, are required to go to 33% by 2020, and to 50% by 2030. Governor Brown, in his January 2015 inaugural address, proposed that the state reach 50% renewable energy as soon as possible as a way to reach United Nations greenhouse gas reduction goals.[274]

In terms of nuclear holdings, the LADWP owns 366 MW (or about 20%) of the Palo Verde nuclear plant. That investment should be sold and the funds used to finance renewable power.

Lower Electric Bills for LA Consumers

Going green can result in lowering the electric bills for consumers if LA invests heavily in energy efficiency. To date, efficiency has "gotten the music," while the money has gone to paying for more and more power plants.

LADWP has the obvious opportunity of offering their customers a package of solar on the roof and efficiency investments so none of that solar power is wasted. With better insulated buildings and more efficient air conditioning and refrigeration powered by the sun, LA citizens can lead the fight against climate change and save money at the same time.

Dave's experience of forty years reveals that most people are too busy trying to make a living to find the time to make the full array of efficiency investments that are cost-effective. Monetary incentives have achieved only a small fraction of what will save money. What is needed is an innovative program. The LADWP should hire "green doctors" who will make "house calls," hire the contractors, inspect the work, and pay for it. It can then require the customer to pay back the investment out of the cost of their energy savings as part of their electricity bill. After the investment is paid back in a few short years, the customer receives the benefit of the saving and a lower energy bill.

Such a program will dramatically reduce the demand for electricity over time and thus make it much easier for LA to shift to an all-electric, all-renewable energy supply.

In addition, such a program will create a sizeable number of new jobs right in the city of LA, where jobs are sorely needed because of the relatively high unemployment rate that has persisted for years.

Making efficient use of renewable energy would end—once and for all—the false notion that we cannot afford to go green. The truth is, we can't afford not to do so.

Motivating the Move to Solar

The most powerful mechanism of change in the ongoing green revolution has been educating consumers about the benefits they stand to gain and incentivizing participation in efficiency and alternative-energy initiatives.

As of this writing, LADWP continues to operate as a monopoly and fails to actively promote solar power on a massive scale or even small-scale development of microgrids owned by customers or competing companies. The city could spur a more rapid transition to renewables by:

- Changing the rate structure so that those who invest in efficiency or build microgrids powered by solar are well compensated
- Standardizing agreements to speed connection to the grid

But more fundamentally, LADWP should start actively selling solar and efficiency to their customers with their low cost of financing and nonprofit status. "Lend Me Your Roof" should be more than a slogan. LADWP could afford to pay a bit for the privilege of putting solar on the roofs of LA. And over time, the whole city would benefit from electricity that has no fuel cost and requires little maintenance.

A Heat Pump for Every Home

To replace the use of natural gas for home heating, the city-owned utility should incentivize the switch to heat pumps. Just as with solar and efficiency programs, the LADWP can install and implement the adoption of heat pumps by their "green doctors."

To get the ball rolling, LADWP could finance heat pumps and allow repayment over time in the customer's energy bills and ultimately save them money. And mind you, the heat pump acts as an air conditioner in the summer. Dave implemented such a program in the Tennessee Valley in 1977 and it is one of his initiatives that still survives today.[275]

Solar Hot Water

Lots of natural gas is used to heat water. Certainly in LA the direct heat of the sun can be the clean substitute. Solar hot water is standard procedure in Israel and the West Bank, where the climate is similar to LA. LADWP could finance solar hot water to its customers and include them in the "green doctor's medicine package." The heat pump systems could be a backup for very cloudy days.

What If You Could Fill Your Tank at the Equivalent of $1 a Gallon?

The widespread adoption of electric cars should be a top priority for LADWP, and every electrical utility. Most utilities have dragged their feet about electric vehicles. But this is a marvelous opportunity to grow their business. Fueling the two million cars in LA at approximately 4,000 KWh per car annually would require a total supply of 8.0 billion kWh, which would add 40% to today's electric use. How many businesses can achieve growth of that kind?[276]

Furthermore, this growth can be managed so as to occur when loads are low at night so that much less new capacity would be needed. The electric system would become more efficient.

Meanwhile, consumers stand to reap considerable savings. The current cost of electricity in LA allows an electric car to be charged at a cost that is the equivalent of less than one dollar for a gallon of gasoline. And the more that electricity comes from renewable sources, and the more cars on the road that are electric, that cost is likely to remain stable. That's not even factoring in how much cleaner the city's air would be.

LA could jump-start the transition to electric cars by taking these actions:

- Decide that LADWP will own the batteries in any electric car purchased by its customers for the next five years. This will reduce the price to consumers, which is a big stumbling block, as well as increase the resale value of the car because the battery is really the main item that wears out in an electric car. The batteries could also be used for storage capacity for the utility. In the smart renewable grid, surplus power is stored in car batteries. These cars can supply part of the storage the smart-grid system needs to be reliable. Offer a low off-peak rate that would be the equivalent of less than one dollar for a gallon of gasoline and still cover the cost to the utility because more expensive power isn't used off-peak. Offer to install a car-charging plug in the home or apartment building for anyone buying an electric car so owners can charge the batteries overnight.
- Provide similar plug-in facilities at all existing car parking places and in other locations throughout the city so that an electric car is never more than five miles from a fast charging station.
- Give discounts to electric cars at city-owned parking lots.
- Continue to give preference to electric cars in the carpool lane just as we do now for cars with two or more passengers.

We are not suggesting that the LADWP go into the automobile business. On the contrary, we are proposing they act like business folks in the electric business. Furthermore, every electric car will provide part of the storage system vital to reliability. It is a win-win for the utility and the customers.

The electric car appears to be the best bet for the near-term and steady reduction of gasoline consumption, but fuel-cell cars and trucks running on hydrogen should also come along to give consumers even more choice of zero-emission motor vehicles.

The Shipping Infrastructure

Great strides can also be made by electrifying the city's infrastructure for the receipt and transport of goods. The Port of LA and the LA Airport (LAX) have both been huge sources of petroleum pollution. In fact, the port has been one of the biggest sources of pollution in the entire region.

But already, this has changed. In 2005, Dave became president of the commission that oversees the Port of LA and with the strong support of then Mayor Villaraigosa, the Port launched a major initiative that has reduced local area pollution by 80%. Soon, all the ships will be powered by electricity when they are tied up at the dock, and all of the huge cranes that move the containers from the ships to the docks will also operate on electricity. But the port won't be truly green until this electricity is made from renewables.

The trucks that serve the port still run on oil and natural gas. The port has actually funded the development of all-electric trucks for short hauls of two hundred miles; it is also continuing to receive redesigned and more efficient short-haul electric trucks and remains committed on the record to advance their use. Clean hydrogen-powered trucks, called "Vision" trucks, have also been developed in partnership with the Port of Long Beach. These are hydrogen-hybrid trucks run on a hydrogen fuel-cell and lithium batteries and have a potential range of

143

up to four hundred miles. Continual leadership is needed to put more zero-emission trucks on the road.[277]

The trains that move the containers long distances to and from the port must also be converted to electricity. An exciting technology to consider is Maglev trains, which are powered by electricity and which are actually levitated above the tracks. They move at high speed and relatively low cost due to the lack of friction between the wheels and track. They're in use in Germany and China already.[278] Another option is a version of a monorail that would move containers on the equivalent of a fast conveyor belt.

Where There's the Will

The implementation of this green vision depends entirely on the priority the city and its people give to doing the hard work required; progress is not constrained by technology. The technology is available. The same is true of the prospects for every municipality and town in the country. The real issue is the will, not the way.

CHAPTER 14

An All-Electric Energy Policy Act

The time is at hand to enact a national energy policy that will put forth a combination of requirements and incentives to achieve both a low-cost energy supply and avoid climate catastrophe. The policy must be comprehensive and long term so that significant progress is made each year over a thirty- to thirty-five-year period.

The proposals now being debated won't solve the problem and keep the U.S. in accord with the 2°C temperature limit. To be effective, the policy must lay out a clear path to an All-Renewable All-Electric energy system, as follows:

1) Outlaw the building of new fossil-fueled electric power plants.

2) Order a steady reduction each year in GHG emissions through these measures:

   a) Require that emissions be reduced 3% each year for the next thirty five years, the amount required to enable the U.S. to reach a nearly GHG-free energy supply by 2050.

   b) Assess a $300 per ton penalty on any amount of GHG emitted above the required reduction, to be enforced by the IRS. Also prohibit offenders from passing on the cost of the penalty to consumers.

c) This law would apply to the electric power industry as well as to the natural gas utilities and major sellers of petroleum for heating.

d) Require that all new homes and buildings be GHG-free and existing buildings be retrofitted to zero GHG at time of sale or within fifteen years.

e) The reductions can be made through a combination of incentivizing the weatherization of buildings, retrofitting of homes with electric heat pumps, installation of solar panels, smart-grid technology, storage systems and the production of GHG-free hydrogen.

3) National Portfolio Standard, requiring every electric utility—both publicly and privately owned—to meet the following requirements:

- 30% GHG-free generation by 2025

- 60% GHG-free generation by 2035

- 100% GHG-free generation by 2050

4) Transform the transportation sector by 2050.

a) For the railroads:

i) Requiring every railroad to draft a detailed plan for converting to electricity within two years and submit it to the Department of Transportation for approval.

ii) Providing government loan guarantees for the financing of the implementation of approved plans.

iii) Requiring that the electrification of the railroads be initiated within ten years and completed within twenty years, or the railroad will be charged a $300/ton GHG fee on all

GHG emitted after that date.

b) For motor vehicles:

    i) Requiring all major auto, truck, and bus manufacturers to reduce GHG emissions of vehicles by 3% each year, through a combination of improvements in mileage and lower GHG emissions.

    ii) Requiring that they transition to zero-emission vehicles according to the same timetable as the energy suppliers—30% of all new sales by 2025, 60% by 2035, and 100% by 2050.

c) For airplanes and ships:

    i) Requiring every airline and ship manufacturing company to draft a detailed plan for converting to hydrogen-powered airplanes and hydrogen-hybrid ships within twenty years and submit it to the Department of Transportation for approval.

5) Create a Federal Green Bank, which provides loan guarantees (not loans) for the financing of railroad electrification and for the construction of renewable electricity power plants that have long-term contracts with electric distribution utilities that are ratepayer-funded.

6) Institute tax credits for zero-GHG emission consumer products.

a) Enact a permanent 20% tax credit for electric heat pumps and investment in energy efficiency in homes and commercial businesses.

b) Enact a five-year 20% tax credit for zero-emission motor vehicles to speed the development of a mass market.

These proposals are designed to inject into the climate debate proposals commensurate with the danger we face both from inaction and from adopting measures that in fact just contribute to the emissions problem. Our hope is that a better-informed public will demand the leadership that presently is so woefully lacking.

A prevailing opinion is that if we could enact a tax on carbon the problem would be solved. But would it really? Even a huge tax on carbon wouldn't assure that all our power plants would be renewable and our cars, homes and other transportation would make the necessary transformation.

The only significant greenhouse gas reductions that have occurred have been when state laws required the electric utilities to increase the percent of their power generated by renewable energy and when we required more miles per gallon from our cars. These advances were not made voluntarily by private industry; they required mandates. That is why, in this chapter, we spell out an array of mandates and incentives that can achieve the emission reductions the scientists say we must.

Mandates have been enacted at the state level and we hope our book inspires the "Green States" to consider expanding mandates to those that we propose. And expectantly, in enough time, the federal government and governments throughout the world will follow.

Updates to the legislative concept, "An All-Electric Energy Policy Act," are available on our website at: http://www.allelectricamerica.com

The Path Ahead
"A Call to Action"

I n 1974 Dave foretold in his book, Energy: The New Era, that, "... man is tampering ignorantly and perhaps dangerously with the planet in a very fundamental way. And if we find that excessive fuel consumption is causing changes in climate, the lead-time for reducing fuel consumption to ward off that threat will be quite short. The effects on a high energy civilization without alternatives could be disastrous."[279] Today, in 2015, this prediction has become a reality. Yet, the future need not be grim. We possess the technology to power all of our energy needs with renewable energy. This is our future. The transformation is underway. However, without action it will not happen fast enough to prevent disastrous climate change.

In this book we have documented some basic truths about the state of the energy industry and climate change. There is good and bad news:

The Bad News

1. The climate scientists agree that the world has at most 35 years to reduce total greenhouse gas (GHG) emissions down to zero,

otherwise we risk severe climate disruption and much of the world becoming virtually uninhabitable.

2. 70% of the GHG emissions are caused by the burning of oil and gas. Their use is increasing, not going down.

3. The controls proposed by President Obama deal only with the 30% emitted by electric power plants and will not decrease the GHG emissions enough to reach the climate goal of keeping the Earth's temperature from rising by two degrees Celsius. Additionally, the controls don't count the increasing methane leakage from greater use of natural gas that offsets the reduction in carbon compared to coal.

4. Any investment in natural gas will stay with us for years to come and result in continued release of carbon dioxide and methane into the air when we should instead be reducing it to zero.

The Good News

1. There exists today commercially available solar and wind power technology and means of storing that renewable power that can AT NO INCREASE in cost replace all of the oil, gas and nuclear power plants, 3% per year down to near zero by 2050.

2. There exists today the technology to substitute renewable electricity for the oil and gas now used in transportation and home heating and for all energy uses. The renewable electricity can be used directly in electric cars, trains, heat pumps, etc. Or it can be converted to hydrogen for use in fuel cell vehicles or to substitute directly for natural gas in industrial uses and new designs for aircraft and ships. We can achieve an All-Renewable All-Electric energy supply by 2050 if we begin to go down that road at once and make progress at about 3% each year.

The United States has been a nation of action throughout our history. In the past we led extraordinary undertakings both at home and on a global scale. When we needed a national highway infrastructure, we built it. When we decided it was necessary to go to the moon, we got there. When attacked by a tyranny that threatened the world, we retooled our nation's industrial infrastructure and in a short five years built tanks, planes and ships to win the Second World War, a task far more challenging than building a renewable energy infrastructure. When faced with serious challenges, the United States has rolled up its sleeves, exercised its ingenuity, found the solutions, and carried out the tasks necessary to get the job done.

Constructing an All-Electric American infrastructure is not a sacrifice, it is an investment that will pay for itself in energy savings, stable energy prices, better health, and in preventing future disasters due to worsening global warming. The choice is clear. A better future is within our reach and the age of fossil fuels must end. The United States must embrace the vision of a truly renewable future and show the world how it can happen. We are on the road to climate hell, but we don't have to be. We can be on a road to climate heaven.

Progress that is meaningful in combating climate change requires laws that will achieve the necessary 3% reduction in all greenhouse gases per year. How can we fail to use the greatest force a civilized society possesses — the rule of law — to get the job done? We suggest utilizing our government to solve a problem that we cannot solve individually; the very reason we have a government.

We see the government as having a critical role because of the limited time we have to make this transformation. However, it is not the government that will ultimately develop and finance the majority of our 21st century energy infrastructure. Investment and entrepreneurship by private industry is essential to our success. Our private industry is already making advancements and we must have more companies follow the example of TESLA and Solar City. In fact, if the energy industry is indeed perceptive, they will see the writing on the wall and join in this transformation.

Let's make the road ahead one of action. We are aware that what we are suggesting may seem to many as seeking a political miracle, but we have no other choice. In fact it doesn't require a miracle, only enough good citizens to agree that we can and must end our use of the poisons that are changing the climate one year at a time.

Endnotes

All websites were accessed on July 26, 2015
unless specified otherwise

PART I: THE PROMISE

Chapter 1: All-Electric Renewable America

1.   Freeberg, Ernest, *Age of Edison*, Penguin Press, 2013, pp 63, 70.

2.   We assume a capacity factor of 4 for a completely renewable electricity infrastructure. The capacity factor for the current electric supply is roughly 2. Converting the entire electricity supply to a renewable system would require double the current capacity (e.g. 2 million MW). Over 35 years this would require roughly 60,000 MW per year. The capacity factor for a completely renewable system is obtained from Table 1 (1,591 GW-2050 total end use load) and from Table 2 (6,447 GW of nameplate capacity) in Jacobson, M.Z. and Delucchi, M.A., "100% clean and renewable wind, water, and sunlight (WWS) all-sector energy roadmaps for the 50 United States," *Energy & Environmental Science, The Royal Society of Chemistry,* May 27, 2015; See also, "Capacity factor," *Wikipedia, "The net capacity factor of a power plant is the ratio of its actual output over a period of time, to its potential output if it were possible for it to operate at full nameplate capacity continuously over the same period of time."* Available at, https://en.wikipedia.org/wiki/Capacity_factor.

3.   Stanway, David, "The US And China Just Made A Landmark Joint Deal On Climate Change," *Business Insider*, Nov 11, 2014, Available

at, http://www.businessinsider.com/r-china-agrees-co2-peak-by-2030-us-to-cut-emissions-by-quarter-2014-11; *See also, * The White House, FACT SHEET: U.S.-China Joint Announcement on Climate Change and Clean Energy Cooperation, *Office of the Press Secretary, * November 11, 2014, Available at, https://www.whitehouse.gov/the-press-office/2014/11/11/fact-sheet-us-china-joint-announcement-climate-change-and-clean-energy-c; *See also,* Haugwitz, Frank, "China 17.8GW, a record solar PV installation target for 2015. Achievable?," *PV Tech,* March 24, 2015, Available at, http://www.pv-tech.org/guest_blog/china_17.8gw_a_record_solar_pv_installation_target_for_2015._achievable; *See also,* Yang, Jianxiang, "China to connect 21.5GW of new capacity in 2015," *Wind Power Monthly,* March 18, 2015, Available at, http://www.windpowermonthly.com/article/1338971/china-connect-215gw-new-capacity-2015.

4.   S. David Freeman, *Winning our Energy Independence, An Energy Insider Shows How,* 45-68, 2007 Gibbs Smith, Publisher; *See also,* Jacobson, M.Z. and Delucchi, M.A., "100% clean and renewable wind, water, and sunlight (WWS) all-sector energy roadmaps for the 50 United States," *Energy & Environmental Science, The Royal Society of Chemistry,* May 27, 2015; *See also,* "100% renewable energy," *Wikipedia,* Available at, https://en.wikipedia.org/wiki/100%25_renewable_energy; *See also,* E.K. Hart and M.Z. Jacobson, "A Monte Carlo Approach to Generator Portfolio Planning and Carbon Emissions Assessments of Systems with Large Penetrations of Variable Renewables," *Renewable Energy,* 36, 2278-2286,doi10.1016/j.renene.2011.01.0152011. Available at http://web.stanford.edu/~ehart/manuscript1_draft/manuscript1_EHart_draft.pdf (2011); *See also,* Jacobson, Mark Z., et al., A low-cost solution to the grid reliability problem with 100% penetration of intermittent wind, water, and solar for all purposes, *Proceedings of the National Academy of Sciences,* publication pending Nov 2015; *See also,* Barney Jeffries, Yvonne Deng, et al., "The Energy Report-100% Renewable Energy by 2050," *WWF, ECOFYS & OMA Technical Report* (2011).

5.   Lopez, Anthony, et al., "U.S. Renewable Energy Technical Potentials: A GIS- Based Analysis," *NREL Technical Report,* NREL/TP-6A20-51946,

July 2012. Available at, http://www.nrel.gov/docs/fy12osti/51946.pdf; *See also*, Jacobson, M.Z. and Delucchi, M.A. "Providing all Global Energy with Wind, Water, and Solar Power, Part I: Technologies, Energy Resources, Quantities and Areas of Infrastructure, and Materials," *Energy Pol.*,39, 1155-1169, <doi:10.1016/j.enpol.2010.11.040> (2011). (Hereinafter, "Jacobson & Delucchi, Part I, 2011"); *See also*, M.Z. Jacobson and M.A. Delucchi, "Providing all global energy with wind, water, and solar power, Part II: Reliability, system and transmission costs, and policies," *Energy Pol.*, 39, 1169, doi:10.1016/j. enpol.2010.11045, (2011). (Hereinafter "Jacobson & Delucchi, Part II, 2011").

6. Jacobson, M.Z. and Delucchi, M.A., "100% clean and renewable wind, water, and sunlight (WWS) all-sector energy roadmaps for the 50 United States," *Energy & Environmental Science, The Royal Society of Chemistry*, May 27, 2015.

7. *Ibid.*

8. *Ibid.* Jacobson et al., found the total nameplate capacity needed in 2050 to be roughly 6.4 TW. To determine that there is roughly 18 times the amount of utility solar PV capacity needed compared to the total 2050 energy capacity required (6.4 TW), we take the difference in capacity factors for each technology into account and adjust the corresponding capacity required for any single technology accordingly; *See also*, Lopez, Anthony, et al., "U.S. Renewable Energy Technical Potentials: A GIS- Based Analysis," *NREL Technical Report*, NREL/TP-6A20-51946, July 2012. We obtained total available capacity potential for utility solar PV from the NREL Tech Report, Available at, http://www.nrel.gov/docs/fy12osti/51946.pdf; *See also*, U.S. Energy Information Administration, "International Energy Outlook 2014," *DOE/EIA-0484*, September 2014. Available at, http:// www.eia.doe.gov/oiaf/ieo/index.html.

9. "PV Facts," *U.S. National Renewable Energy Laboratory (NREL)*, Feb 2004, Available at, http://www.nrel.gov/docs/fy04osti/35097. pdf (Accessed March 4, 2015); *See Also*, Lopez, Anthony, et al., "U.S. Renewable Energy Technical Potentials," *NREL*, July 2012.

10. Lopez, Anthony, et al., "U.S. Renewable Energy Technical Potentials," *NREL,* July 2012. Offshore wind capacity is underestimated and there may actually be more than two times the 2050 wind capacity. The NREL study states that, *"Because no offshore or near-shore estimates were available for Florida or Alaska (at the time of this publication), these states are omitted from the technical potential calculations."*

11. IRENA, "Renewable Power Generation Costs in 2014," *International Renewable Energy Agency (IRENA) Report,* January 2015; *See also,* Gore, Al, "The Turning Point: New Hope for the Climate, It's time to accelerate the shift toward a low-carbon future," *Rolling Stones,* June 18, 2014, Available at, http://www.rollingstone.com/politics/news/the-turning-point-new-hope-for-the-climate-20140618; *See also,* Jacobson, M.Z. and Delucchi, M.A., "100% clean and renewable wind, water, and sunlight (WWS) all-sector energy roadmaps for the 50 United States," *Energy & Environmental Science, The Royal Society of Chemistry,* May 27, 2015; *See also,* Lazard, "Lazard Levelized Cost of Energy Analysis-Version 8.0," *Lazard.com,* September 2014, Available at: http://www.lazard.com/media/1777/levelized_cost_of_energy_-_version_80.pdf; *See also,* US EIA, "Levelized Cost and Levelized Avoided Cost of New Generation Resources in the Annual Energy Outlook 2014," *US Energy Information Administration,* April 2014, Available at, http://www.eia.gov/forecasts/aeo/pdf/electricity_generation.pdf. From these resources it is found that the current levelized cost of wind energy is roughly between 4 to 10 cents per kWh; *See also,* Parkinson, Giles, "Solar Costs Will Fall Another 40% In 2 Years. Here's Why," *CleanTechnica,* January 29, 2015, Available at http://cleantechnica.com/2015/01/29/solar-costs-will-fall-40-next-2-years-heres/.

12. *Ibid; See also,* LaMonica, Martin, "Xcel Energy Buying Utility-Scale Solar at Prices Competitive With Natural Gas," *GreenTech Media,* October 2, 2013. Available at, http://www.greentechmedia.com/articles/read/xcel-energy-buys-utility-scale-solar-for-less-than-natural-gas; *See also,* Jacobs, Mike, "Where Is Wind Energy Cheaper than Natural Gas?" *Union of Concerned Scientists,* October 18, 2013, http://blog.ucsusa.org/where-is-wind-energy-cheaper-than-natural-gas-276; *See also,* Irwin, Conway, "Midwest Wind Cost-Competitive With

Gas and Coal," *GreenTech Media*, December 2013, *"In the Midwest, we're now seeing power agreements being signed with wind farms at as low as $25 per megawatt-hour,"* said Stephen Byrd, Morgan Stanley's Head of North American Equity Research for Power & Utilities and Clean Energy, at the Columbia Energy Symposium in late November. *"Compare that to the variable cost of a gas plant at $30 per megawatt-hour. The all-in cost to justify the construction of a new gas plant would be above $60 per megawatt-hour."* Available at, http://www.greentechmedia.com/articles/read/midwest-wind-cost-competitive-with-gas-and-coal. *See also,* http://blog.ucsusa.org/where-is-wind-energy-cheaper-than-natural-gas-276.

13. IRENA, "Renewable Power Generation Costs in 2014," *International Renewable Energy Agency (IRENA) Report*, January 2015; *See also, National Resource Defense Council,* Available at, http://www.nrdc. org/energy/renewables/wind.asp, Costs of onshore-wind are expected to fall to a point where they eventually become consistently lower in different geographic areas, such as low-wind zones, than the cheapest traditional energy sources, such as natural gas; *See also,* "Wind power in the United States," *Wikipedia,* http://en.wikipedia. org/wiki/Wind_power_in_the_United_States.

14. Lopez, Anthony, et al., "U.S. Renewable Energy Technical Potentials," *NREL,* July 2012.

15. Phillips, Ari, "New Report: Low Oil Prices Won't Hurt Clean Energy," *Think Progress,* January 20, 2015, Available at, http://thinkprogress. org/climate/2015/01/20/3613176/low-oil-prices-clean-energy/.

16. IRENA, "Renewable Power Generation Costs in 2014," *International Renewable Energy Agency (IRENA) Report*, January 2015. Available at, http://www.irena.org/publications; *See also,* Cardwell, Dianne, "Solar and Wind Energy Start to Win on Price vs. Conventional Fuels," *New York Times,* Nov 23 2014, Available at, http://www.nytimes. com/2014/11/24/business/energy-environment/solar-and-wind-energy-start-to-win-on-price-vs-conventional-fuels.html?_r=0.

17. LaMonica, Martin, "Xcel Energy Buying Utility-Scale Solar at Prices Competitive With Natural Gas," *GreenTech Media,* October 2, 2013.

Available at, http://www.greentechmedia.com/articles/read/xcel-
energy-buys-utility-scale-solar-for-less-than-natural-gas; *This is the
first time that we've seen, purely on a price basis, that the solar projects made the cut—without considering carbon costs or the need to
comply with a renewable energy standard—strictly on an economic
basis," David Eves, CEO of an Xcel subsidiary, told the Denver Business
Journal; See also,* Proctor, Cathy, "Xcel Energy hopes to triple Colorado
solar, add wind power," *Denver Business Journal,* September 9, 2013
Available at, http://www.bizjournals.com/denver/blog/earth_to_
power/2013/09/xcel-energy-proposes-to-triple-solar.html?page=all;
See also, Shahan, Zachary, "Solar Less Than 5¢/kWh In Austin,
Texas! (Cheaper Than Natural Gas, Coal, & Nuclear)," *Clean Technica,*
March 13, 2014, Available at, http://cleantechnica.com/2014/03/13/
solar-sold-less-5¢kwh-austin-texas/.

18.   Bronski, Peter, Lilienthal ,Peter, and Crowdis, Mark, et
al., "The Economics of Grid Defection When and Where
distributed Soar Generation Plus Storage Competes
with Traditional Utility Service," *Rocky Mountain
Institute,* February 2014, Available at http://www.rmi.org;
See also, Wile, Rob, "GOLDMAN: Solar Is On The Way To Dominating
The Electricity Market, And The World Has Elon Musk To Thank,"
Business Insider, March 18, 2014. Available at, http://www.busi-
nessinsider.com/goldman-on-solar-and-elon-musk-2014-3; *See also,*
Romm, Joe, "Must-See Chart: Cost Of PV Cells Has Dropped An
Amazing 99% Since 1977, Bringing Solar Power To Grid Parity,"
Think Progress, October 6, 2013, Available at, http://thinkprogress.
org/climate/2013/10/06/2717791/cost-pv-cells-solar-power-grid-parity/;
See also, Romm Joe, "Solar Power Is Now Just As Cheap As
Conventional Electricity In Italy And Germany," March 24, 2014,
Available at, http://thinkprogress.org/climate/2014/03/24/3418145/
solar-grid-parity-italy-germany/.

19.   IRENA, "Renewable Power Generation Costs in 2014," *International
Renewable Energy Agency (IRENA) Report,* January 2015. Available at,
www.irena.org/publications; *See also,* Cardwell, Dianne, "Solar and
Wind Energy Start to Win on Price vs. Conventional Fuels, *New York*

Times, Nov 23 2014, Available at, http://www.nytimes.com/2014/11/24/
business/energy-environment/solar-and-wind-energy-start-to-win-
on-price-vs-conventional-fuels.html?_r=0; *See also*, Lazard, "Lazard
Levelized Cost of Energy Analysis-Version 8.0," September 2014; *See
also*, Jacobson, M.Z. and Delucchi, M.A., "100% clean and renewable
wind, water, and sunlight (WWS) all-sector energy roadmaps for the
50 United States," *Energy & Environmental Science, The Royal Society
of Chemistry*, May 27, 2015.

20. *Ibid.*

21. *Ibid.*

22. Gore, Al, "The Turning Point: New Hope for the Climate, It's time
to accelerate the shift toward a low-carbon future," *Rolling Stone*,
June 18, 2014.

23. Andrea Reimer, "100% Renewable Energy: The new normal?"
Huffington Post, April 24, 2014. Available at: http://www.
huffingtonpost.com/andrea-reimer/100-renewable-energy-the-new-
normal_b_7126906.html.

24. Justin Gillis, "A Tricky Transition From Fossil Fuel Denmark Aims
for 100 Percent Renewable Energy," *New York Times*, Nov 10, 1014.
Available at: http://www.nytimes.com/2014/11/11/science/earth/
denmark-aims-for-100-percent-renewable-energy.html?_r=0; *See also*,
James Cave, "Hawaii Wants To Be The First State To Run Completely
On Renewable Energy," *Huffington Post*, April 24, 2015. Available at:
http://www.huffingtonpost.com/2015/04/24/hawaii-renewable-en-
ergy_n_7132844.html; *See also*, Richardson, Jake, "100% Renewable
Energy Goal For Hawaii: Governor Signs Bill," *CleanTechnica*, June 11,
2015, Available at, http://cleantechnica.com/2015/06/11/100-renew-
able-energy-goal-hawaii-governor-signs-bill/.

25. Leahy, Stephen, "Vancouver commits to run on 100%
renewable energy," *The Guardian*, April 2015, Available at:
http://www.theguardian.com/environment/2015/apr/10/
vancouver-commits-to-run-on-100-renewable-energy.

26. COMMITTEE ON ENERGY AND COMMERCE, ONE HUNDRED

THIRTEENTH CONGRESS, Memorandum, September 17, 2013 http://democrats.energycommerce.house.gov/sites/default/files/documents/Memo-EP-Climate-Change-Obama-Administration-2013-9-17.pdf; *See also*, http://change.gov/agenda/energy_and_environment_agenda/. http://change.gov/agenda/energy_and_environment_agenda/

27. US Department of Energy, "All Electric Vehicles," Available at, https://www.fueleconomy.gov/feg/evtech.shtml.

28. Jacobson, M.Z. and Delucchi, M.A., "100% clean and renewable wind, water, and sunlight (WWS) all-sector energy roadmaps for the 50 United States," *Energy & Environmental Science, The Royal Society of Chemistry*, May 27, 2015. *"Conversion would reduce each state's end-use power demand by a mean of ~39.3% with ~82.4% of this due to the efficiency of electrification and the rest due to end-use energy efficiency improvements."* This is a conservative number. They find energy use is less because of increased efficiency from factors such as electric car and heaters in homes being more efficient than combustion.

29. "US Gross Domestic Product GDP History," Available at, http://www.usgovernmentspending.com/us_gdp_history.

30. Index mundi, United States—GDP, Available at, http://www.index-mundi.com/facts/united-states/gdp.

31. NREL, "Hydrogen Basics," *National Renewable Energy Labs*, Available at, http://www.nrel.gov/learning/eds_hydrogen.html.

32. Holland, Geoffrey B. and Provenzado, James J., T*he Hydrogen Age, Empowering a Clean Energy Future*, Gibbs Smith Publishers, 2007.

33. NREL, "Hydrogen and Fuel Cell Research, *National Renewable Energy Labs*, Available at, http://www.nrel.gov/hydrogen/proj_production_delivery.html; *See also*, Holland, Geoffrey B. and Provenzado, James J., *The Hydrogen Age, Empowering a Clean Energy Future*, Gibbs Smith Publishers, 2007.

34. Warrick, Joby, "Utilities wage campaign against rooftop solar," *The Washington Post*, March 7, 2015, Available at, http://www.washingtonpost.com/national/health-science/

utilities-sensing-threat-put-squeeze-on-booming-solar-roof-indus-
try/2015/03/07/2d916f88-c1c9-11e4-ad5c-3b8ce89f1b89_story.html.

35. Parks, L.Y., "Storage: The Holy Grail Liveth," *ElectricityPolicy.
com*, October 15, 2013; *See also*, Dragoon, Ken, "Energy Storage
Opportunities and Challenges, prepared for EDF Renewable Energy,"
Ecofys, April 4, 2014; *See also*, Akhil, Abbas A., et al., "DOE/EPRI 2013
Electricity Storage Handbook in Collaboration with NRECA," *Sandia
Report*, Sandia National Laboratories, SAND2013-5131, July 2013,
Available at, http://www.sandia.gov/ess/publications/SAND2013-5131.
pdf.

36. Peterman, Hon. Carla J., and Charles, Melicia, "Tapping Into the
Potential of Energy storage," *ElectricityPolicy.com*, March 5, 2014;
See also, California Energy Storage Bill, Assembly Bill No. 2514,
Available at, http://leginfo.legislature.ca.gov/faces/billNavClient.
xhtml?bill_id=200920100AB2514; *See also*, Marritz, Robert, "New York
regulators' process and vision, still being crafted, merits attention,"
ElectricityPolicy.com, Available at, http://www.electricitypolicy.com/
Editorials/new-york-regulators-process-and-vision-still-being-craft-
ed-merits-attention; *See also*, New York State, "14-M-0101: Reforming
the Energy Vision (REV)," *NY.Gov*, Available at, http://www3.dps.
ny.gov/W/PSCWeb.nsf/All/26BE8A93967E604785257CC40066B91A.

37. Wile, Rob, "GOLDMAN: Solar Is On The Way To Dominating The
Electricity Market, And The World Has Elon Musk To Thank," *Business
Insider*, March 18, 2014, Available at, http://www.businessinsider.com/
goldman-on-solar-and-elon-musk-2014-3; *See also*, Smith, Noah, "Clean
Energy Revolution Is Ahead of Schedule," *Bloomberg View*, April 8 2015,
Available at, http://www.bloombergview.com/articles/2015-04-08/
clean-energy-revolution-is-way-ahead-of-schedule; *See also*, Guest
Contributor, "Energy Storage Could Reach Big Breakthrough Price
Within 5 Years," *CleanTechnica*, March 4, 2015, Available at, http://
cleantechnica.com/2015/03/04/energy-storage-could-reach-cost-
holy-grail-within-5-years/; *See also*, Navigant Research, "Global
Energy Storage Installations: Market Share Data, Industry Trends,
Market Analysis, and Project Tracking by World Region, Technology,
Application, and Market Segment," *Energy Storage Tracker 1Q15*,

Available at, https://www.navigantresearch.com/research/energy-stor-age-tracker-1q15; *See also*, Shankleman, Jessica, "Barclays down-grades US power sector over solar threat," *Business Green*, May 30, 2014, Available at, http://www.businessgreen.com/bg/news/2347374/barclays-downgrades-us-power-sector-over-solar-threat.

38.   Doug, Parr, "Comment: The solar storage energy revolution is arriving," *Energy Desk, GreenPeace*, April 27, 2015, Available at, http://energydesk.greenpeace.org/2015/04/27/comment-the-solar-storage-en-ergy-revolution-is-arriving/; *See also*, Ferrell, John, "The Next Charge for Distributed Energy," *Institute for Local Self-Reliance (ILSR)*, March 2014; *See also*, Carbajales-Dale, M., et al., "Can we afford storage? A dynamic net energy analysis of renewable electricity generation supported by energy storage," *Energy Environ. Sci.*, February 5, 2014, 7, 1538–1544; *See also*, Bronski, Peter, Lilienthal, Peter, and Crowdis, Mark, et al., "The Economics of Grid Defection When and Where distributed Soar Generation Plus Storage Competes with Traditional Utility Service," *Rocky Mountain Institute*, February 2014, Available at http://www.rmi.org.

39.   Denning, Liam, "Tesla's Battery Could Power Utilities Elon Musk's new battery packs are a threat to utilities, but they also repre-sent an opportunity to maintain relevance," *WSJ*, May 1st, 2015, http://www.wsj.com/articles/teslas-battery-could-power-utili-ties-1430505757; *See also*, McMahon, Jeff, "Why Tesla Batteries Are Cheap Enough To Prevent New Power Plants," *Forbes*, May 5, 2015, Available at, http://www.forbes.com/sites/jeffmcmahon/2015/05/05/why-tesla-batteries-are-cheap-enough-to-prevent-new-power-plants/.

40.   "Natural Gas Consumers," *EIA*, Available at, http://www.eia.gov/dnav/ng/ng_cons_num_a_epg0_vn3_count_a.htm & http://www.eia.gov/dnav/ng/ng_cons_num_a_EPG0_VN7_Count_a.htm.

41.   Larsen, Kate, et al., "Untapped Potential, Reducing Global Methane Emissions from Oil and Natural Gas Systems," *Rhodium Group Report*, April 2015; *See also*, Busch, Chris & Gimon, Eric, "Natural Gas versus Coal," *The Electricity Journal*, Aug/Sept 2014; *See also*, Howarth R.W., "A bridge to nowhere: methane emissions and the greenhouse gas

ENDNOTES

footprint of natural gas," *Energy Science & Engineering*, Vol. 2 no. 2, pp. 47–60, June 2014; *See also*, Allen, David T., et al., "Measurements of methane emissions at natural gas production sites in the United States," PNAS, Vol. 110 no. 44, pp. 17768–17773, October 29, 2013; *See also*, EPA, "Inventory of U.S. Greenhouse Gas Emissions and Sinks: 1990-2011, *US EPA* ,430-R-13-001, April 2013; *See also*, EPA, "Inventory of U.S. Greenhouse Gas Emissions and Sinks: 1990-2012," *U.S. Environmental Protection Agency Report,* April 15, 2015, Often referred to in this chapter as, EPA 2015 Inventory, Available at, http://www3.epa.gov/climatechange/Downloads/ghgemissions/US-GHG-Inventory-2015-Main-Text.pdf, see Chapter 7.

42.   IPCC (2013), *Climate Change 2013: The Physical Science Basis. Contribution of Working Group I to the Fifth Assessment Report of the Intergovernmental Panel on Climate Change,* Cambridge University Press, Cambridge, United Kingdom and New York, NY, USA, 1535 (Hereinafter referred to in this chapter, IPCC (2013), WGI, AR5). The IPCC 2013 determined a GWP for methane compared to CO_2 to be 84 times larger. Other studies have found higher numbers at 105.

43.   Busch, Chris & Gimon, Eric, "Natural Gas versus Coal: Is Natural Gas Better for the Climate? *The Electricity Journal*, doi;10.1016/j.tej.2014.07.007, August 2014*; See also*, Wigley, T. M. L., "Coal to gas: the influence of methane leakage," *Climatic Change* 108: 601–608, 2011, See chapter 7 for a detailed analysis.

44.   Lawrence Livermore National Labs, "Estimated U.S. Carbon Dioxide Emission in 2013: 5390 Million Metric Tons," Available at, https://flowcharts.llnl.gov/carbon.html#2013.

45.   See chapter 7 for a detailed analysis.

46.   IPCC (2013), WGI, AR5, TS-61,27, 28, 103, 485; *See also*, Freedman, Andrew, "IPCC Report Contains 'Grave' Carbon Budget Message," *Climate Central*, October 4, 2013, Available at, http://www.climate-central.org/news/ipcc-climate-change-report-contains-grave-car-bon-budget-message-16569; *See also*, "Carbon Dioxide Information Analysis Center," http://cdiac.esd.ornl.gov.

47.   "Copenhagen Accord," *Wikipedia*, Available at, http://en.wikipedia. org/wiki/Copenhagen_Accord.

48.   IPCC (2013), WGI, AR5, TS-61, 27, 28, 103, 485, The United Nation's International Panel of Climate Control IPCC has calculated that the maximum amount of carbon (carbon equivalents) from greenhouse gases (budget) that the atmosphere can safely hold beyond its preindustrial level through 2100, without serious disruption to the environment, amounts to 790 gigatons of carbon GtC. We have already emitted roughly 520 GtC leaving only 270 GtC left to burn safely between 2012–2100. See Chapter 8 for a more detailed analysis.

49.   "Carbon dioxide equivalent," *Wikipedia*, Available at, http://en.wiki-pedia.org/wiki/Carbon_dioxide_equivalent.

50.   Based on our calculations, the only way to not emit more than 270 GtC_{eq} (the remaining carbon equivalent budget) by the end of the century is for the world to reach zero greenhouse gas emissions by midcentury; *See also*, Anderson, Kevin and Bows, Alice, "Beyond 'dangerous' climate change: emission scenarios for a new world," *Phil. Trans. R. Soc. A,* 369, 20–44 doi:10.1098/rsta.2010.0290, 2011; *See also*, Paltsev, Sergey et al., "What GHG Concentration Targets are Reachable in this Century?" *MIT Joint Program on the Science and Policy of Global Change*, Report No. 247 July 2013. See Chapter 8 for a more detailed analysis.

51.   IPCC, "Working Group III Mitigation, Historic Trends and Driving Forces," Available at, http://www.ipcc.ch/ipccreports/tar/wg3/ index.php?idp=125; *See also*, McGlade, Christophe Ekins, Paul, "The geographical distribution of fossil fuels unused when limiting global warming to 2 °C," *Nature*, 517, 187–190, January 8, 2015, Available at, http://www.nature.com/nature/journal/v517/n7533/full/nature14016. html.

52.   Romm, Joe, "Future Gen Dead again: Obama Pulls Plugg and Never Gen Clean Coal Project," *Think Progress*, February 2015. http://think-progress.org/climate/2015/02/05/3619195/futuregen-clean-coal-proj-ect-dead/: In fact, because of such low energy-to-cost ratios, Obama cancelled the FutureGen 2.0 project which was a failed Bush-era

"clean coal" project. See Chapter 8 for additional references.

Chapter 2: Solar and Wind Can Electrify Everything

53. Sunshot U.S. Department of Energy, "SunShot Vision Study," *U.S. DOE Report*, February 2012, Available at: http://www1.eere.energy. gov/solar/pdfs/47927.pdf.

54. Lopez, Anthony, et al., "U.S. Renewable Energy Technical Potentials: A GIS-Based Analysis," *NREL Technical Report*, NREL/TP-6A20-51946, July 2012, Available at, http://www.nrel.gov/docs/fy12osti/51946.pdf; *See also*, Jacobson, M.Z. and Delucchi, M.A. "Providing all Global Energy with Wind, Water, and Solar Power, Part I: Technologies, Energy Resources, Quantities and Areas of Infrastructure, and Materials," *Energy Pol.*, Vol. 39, pp. 1155–1169, 2011; *See also*, Jacobson, M.Z. and Delucchi, M.A., "Providing all global energy with wind, water, and solar power, Part II: Reliability, system and transmission costs, and policies," *Energy Pol.*, Vol. 39, pp. 1169, 2011; *See also*, Barney Jeffries, Yvonne Deng, et al., "The Energy Report-100% Renewable Energy by 2050," *WWF, ECOFYS & OMA Technical Report* (2011).

55. Lawrence Livermore National Laboratory, "Estimated US Energy use in 2014: 98.3 Quads," available at: https://flowcharts.llnl.gov; *See also*, U.S. Energy Information Administration http://www.eia.gov/ forecasts/steo/report/renew_co2.cfm; *See also*, EIA, "Table 1.1.A. Net Generation from Renewable Sources: Total (All Sectors), 2005–March 2015," *Electric Power Monthly*, Available at, http://www.eia.gov/elec-tricity/monthly/epm_table_grapher.cfm?t=epmt_1_1_a, *See also*, EIA, "Table 1.1. Net Generation by Energy Source: Total (All Sectors), 2005–March 2015," *Electric Power Monthly*, Available at, http://www. eia.gov/electricity/monthly/epm_table_grapher.cfm?t=epmt_1_1. *See also*, http://en.wikipedia.org/wiki/Solar_power_in_the_United_States.

56. Jacobson and Delucchi II & II, 2011; *See also*, Jacobson M.Z. & Delucchi M.A, et al., "100% clean and renewable wind, water, and sunlight (WWS) all-sector energy roadmaps for the 50 United States," *Energy and Environmental Science*, May 2015; *See also*, Cory Budischack et al., "Cost-minimized combinations of wind power,

solar power and electrochemical storage, powering the grid up to 99.9% of the time," *J. POWER SOURCES*, Vol. 225, pp. 60–74, March 1, 2013, published findings that wind and solar energy alone combined with storage are capable of powering 99.9% of the energy required by the eastern United States. They found the cost similar to today when considering 2030 technology; *See also*, IRENA, "Renewable Power Generation Costs," *IRENA Report,* January 2015, Available at, http://www.irena.org/DocumentDownloads/Publications/IRENA_ RE_Power_Costs_2014_report.pdf.

57. Wile, Rob, "GOLDMAN: Solar Is On The Way To Dominating The Electricity Market, And The World Has Elon Musk To Thank," *Business Insider*, March 18, 2014, Available at, http://www.businessinsider. com/goldman-on-solar-and-elon-musk-2014-3; *See also*, Parkinson, Giles, "Solar Costs Will Fall Another 40% In 2 Years. Here's Why," *CleanTechnica*, January 29, 2015, Available at http://cleantechnica. com/2015/01/29/solar-costs-will-fall-40-next-2-years-heres/.

58. Parkinson, "Giles, Solar Costs Will Fall Another 40% In 2 Years. Here's Why," *CleanTechnica*, January 29, 2015, Available at http:// cleantechnica.com/2015/01/29/solar-costs-will-fall-40-next-2-years- heres/; *See also*, Mayer, Johannes N., Current, "Current and Future Cost of Photovoltaics, Long-term Scenarios for Market Development, System Prices and LCOE of Utility-Scale PV Systems," *Fraunhofer- Institute for Solar Energy Systems (ISE)*, Study on behalf of Agora Energiewende, February 2015; *See also*, IRENA, "Renewable Power Generation Costs, *IRENA Report*, January 2015, Available at, http:// www.irena.org/DocumentDownloads/Publications/IRENA_RE_ Power_Costs_2014_report.pdf.

59. Wesoff, Eric, "First Solar CEO: 'By 2017, We'll Be Under $1.00 per Watt Fully Installed," *GreenTech Media*, June 24, 2015, Available at, http://www.greentechmedia.com/articles/read/ First-Solar-CEO-By-2017-Well-be-Under-1.00-Per-Watt-Fully-Installed.

60. Lacey, Stephen, "Cheapest Solar Ever: Austin Energy Gets 1.2 Gigawatts of Solar Bids for Less Than 4 Cents," *GreenTech Media*, June, 30 2015, Available at, https://www.greentechmedia.com/articles/

read/cheapest-solar-ever-austin-energy-gets-1.2-gigawatts-of-solar-bids-for-less.

61. Proctor, Cathy, "Xcel Energy hopes to triple Colorado solar, add wind power," *Denver Business Journal*, September 9, 2013, Available at, http://www.bizjournals.com/denver/blog/earth_to_power/2013/09/xcel-energy-proposes-to-triple-solar.html?page=all.

62. Hulac, Benjamin and ClimateWire, "Tesla's Elon Musk Unveils Solar Batteries for Homes and Small Businesses," *Scientific American,* May 1, 2015, Available at, http://www.scientificamerican.com/article/tesla-s-elon-musk-unveils-solar-batteries-for-homes-and-small-businesses/.

63. Lopez, Anthony, et al., "U.S. Renewable Energy Technical Potentials: A GIS-Based Analysis," *NREL Technical Report,* July 2012, NREL/TP-6A20-51946, Available at http://www.osti.gov/bridge.

64. Wile, Rob, "Barclays Has the Best Explanation of How Solar Will Destroy American Utilities," *Business Insider*, May, 28, 2015, Available at, http://www.businessinsider.com/barclays-downgrades-utilities-on-solar-threat-2014-5.

65. Hallock, Lindsey & Rob, Sargent, Shining Rewards, "The Value of Rooftop Solar Power for Consumers and Society," *Environment America Research & Policy Center & Frontier Group,* Summer 2015, Available at, http://www.environmentamerica.org/sites/environment/files/reports/EA_shiningrewards_print.pdf; *See also*, Barney Jeffries, Yvonne Deng, et al., "The Energy Report-100% Renewable Energy by 2050," *WWF, ECOFYS & OMA Technical Report* (2011).

66. Jacobson M.Z. & Delucchi M.A, et al., "100% clean and renewable wind, water, and sunlight (WWS) all-sector energy roadmaps for the 50 United States," *Energy and Environmental Science*, May 2015.

67. Gore, Al, "The Turning Point: New Hope for the Climate, It's time to accelerate the shift toward a low-carbon future," *Rolling Stone,* June 18, 2014; available at, http://www.rollingstone.com/politics/news/the-turning-point-new-hope-for-the-climate-20140618. *See also,* http://www.rollingstone.com/politics/news/the-turning-point-new-hope-for-the-climate-20140618.

68. American Wind Energy Association, Available at http://www.awea. org/Resources/Content.aspx?ItemNumber=5059.

69. EIA Electric Power Monthly, *EIA,* Available at, http://www.eia.gov/ electricity/monthly/epm_table_grapher.cfm?t=epmt_1_1_a & http:// www.eia.gov/electricity/monthly/epm_table_grapher.cfm?t=epmt_1_1.

70. Lopez, Anthony, et al., "U.S. Renewable Energy Technical Potentials: A GIS-Based Analysis," July 2012, *NREL Technical Report,* NREL/ TP-6A20-51946, We obtained total available capacity potential for wind (on-shore and off-shore) from the NREL Tech Report, Available electronically at http://www.nrel.gov/docs/fy12osti/51946.pdf; *See also,* Jacobson M.Z. & Delucchi M.A, et al., "100% clean and renewable wind, water, and sunlight (WWS) all-sector energy roadmaps for the 50 United States," *Energy and Environmental Science,* May 2015. Jacobson et al. found the nameplate capacity needed in 2050 to be roughly 6.4 TW. To determine that there is roughly 2.5 times the amount of wind capacity needed compared to the total 2050 energy capacity required (6.4 TW), we take the difference in capacity factors for each technology into account and adjust the corresponding capacity required for any single technology accordingly.

71. Gore, Al, "The Turning Point: New Hope for the Climate, It's time to accelerate the shift toward a low-carbon future," *Rolling Stones,* June 18, 2014, Available at, http://www.rollingstone.com/politics/news/ the-turning-point-new-hope-for-the-climate-20140618

72. "The ISO – How Renewable Energy Can Save Ratepayer Money," *Conservation Law Foundation,* Available at, http://www.clf.org/blog/ clean-energy-climate-change/renewable-energy-saves-money/

73. Stromsta , Karl-Erik, "2015 could be breakthrough year for US renew-ables policy, ACORE says," *Recharge,* December 2014, Available at, http://www.rechargenews.com/wind/1387488/2015-could-be-break-through-year-for-US-renewables-policy-ACORE-says.

74. LaMonica, Martin, "Xcel Energy Buying Utility-Scale Solar at Prices Competitive With Natural Gas," *GreenTech Media,* October 2, 2013, Available at, http://www.greentechmedia.com/articles/read/

xcel-energy-buys-utility-scale-solar-for-less-than-natural-gas; *See also,* Jacobs, Michael, "Where Is Wind Energy Cheaper than Natural Gas?" *Union of Concerned Scientists,* October 18, 2013, Available at, http://blog.ucsusa.org/where-is-wind-energy-cheaper-than-natural-gas-276.

75. Irwin, Conway, "Midwest Wind Cost-Competitive With Gas and Coal," *Greentech Media,* December 7, 2013, *"In the Midwest, we're now seeing power agreements being signed with wind farms at as low as $25 per megawatt-hour,"* said Stephen Byrd, Morgan Stanley's Head of North American Equity Research for Power & Utilities and Clean Energy, at the Columbia Energy Symposium in late November. *"Compare that to the variable cost of a gas plant at $30 per megawatt-hour. The all-in cost to justify the construction of a new gas plant would be above $60 per megawatt-hour."* Available at, http://www.greentechmedia.com/articles/read/midwest-wind-cost-competitive-with-gas-and-coal.

76. Federal Production Tax Credit for wind energy, *AWEA,* Available at, http://www.awea.org/Advocacy/Content.aspx?ItemNumber=797.

77. "Will Congress extend solar tax incentives expiring in 2016," *GW Solar Institute,* Available at, http://solar.gwu.edu/q-a/will-congress-extend-solar-tax-incentives-expiring-2016.

78. "Birds and Collisions," *American Bird Conservancy,* Available at, http://web.archive.org/web/20150721125312/http://www.abcbirds.org/abcprograms/policy/collisions/index.html; *See also,* Jacobson, M.Z. "Myths and Realities about Wind, Water, and Sun (WWS) Versus Current Fuels," September 26, 2012.

79. Sovacool, Benjamin, "The Avian and Wildlife Costs of Fossil Fuels and Nuclear Power," *Journal of Integrative Environmental Sciences,* Vol. 9, No. 4, December 2012, 255–278; *See also,* Jacobson, M.Z. "Myths and Realities about Wind, Water, and Sun (WWS) Versus Current Fuels," September 26, 2012.

80. Jacobson and Delucchi, "100% all-sector energy roadmaps for the 50 United States," 2015; *See also,* Lovins, Amory B., "Renewable Energy's "footprint" myth," *The Electricity Journal,* Vol. 24 No.6, pp. 40–47, June 2011.

81. "Offshore Wind and Wildlife," *National Wildlife Federation,* Available at, http://www.nwf.org/What-We-Do/Energy-and-Climate/Renewable-Energy/Offshore-Wind/Offshore-Wind-Wildlife-Impacts.aspx; *See also,* "Danish Offshore Wind-Key Environmental Issues," *DONG Energy, Vattenfall, The Danish Energy Authority and The Danish Forest and Nature Agency,* November 2006; *See also,* University of Maryland Center for Environmental science, "New Study Calls for continuing impacts of offshore wind farms on marine species," Available at, http://www.umces.edu/cbl/release/2014/oct/13/assess-impacts-off-shore-wind-farms-marine-specie. It is true that studies come out with concerns that we must take seriously, but we are running out of time and must act immediately as we may lose more marine life to global warming than to installation offshore wind. As stated by the National Wildlife Federation, *"Like any energy development—if done without proper planning, siting, risk assessment and design, there is a potential for offshore wind to negatively affect wildlife ... While conditions differ in Europe, offshore wind energy has been developed extensively there, and studies have found no significant or long-term impacts on wildlife in the area ... Offshore wind energy is a critical part in cutting carbon pollution and reducing the impact of climate change on wildlife. According to the world's leading scientists, as many as 30% of species worldwide will face extinction this century if warming trends continue. To protect wildlife from the dangers of a warming world, we must take appropriate, responsible action to replace as much of our dirty fossil fuel use with clean renewable energy sources. And wind is a key part of that task."*

82. Jacobson and Delucchi, 100% all-sector energy roadmaps for the 50 United States, 2015.

83. *Ibid.* The other renewable sources are ~1.25% geothermal power, ~0.37% wave power, ~0.14% tidal power, and ~3.01% hydroelectric power. *"Based on a parallel grid integration study, an additional ~4.4% and ~7.2% of power beyond that needed for annual loads would be supplied by CSP with storage and solar thermal for heat, respectively, for peaking and grid stability. Over all 50 states, converting would provide ~3.9 million 40-year construction jobs and ~2.0 million 40-year*

operation jobs for the energy facilities alone, the sum of which would outweigh the ~3.9 million jobs lost in the conventional energy sector. Converting would also eliminate ~62000 (19000–115000) U.S. air pollution premature mortalities per year today and ~46000 (12000–104000) in 2050, avoiding ~$600 ($85–$2400) bil. per year (2013 dollars) in 2050, equivalent to ~3.6 (0.5–14.3) percent of the 2014 U.S. gross domestic product. Converting would further eliminate ~$3.3 (1.9–7.1) tril. per year in 2050 global warming costs to the world due to U.S. emissions. These plans will result in each person in the U.S. in 2050 saving ~$260 (190–320) per year in energy costs ($2013 dollars) and U.S. health and global climate costs per person decreasing by ~$1500 (210–6000) per year and ~$8300 (4700–17600) per year, respectively." A state-by-state analysis can be found at http://www.thesolutionsproject.org.

Chapter 3: The Energy Storage Solution

84.   Deign, Jason, "Abengoa: Could molten salt do peaker job?" *Energy Storage Report*, July 1, 2015, Available at, http://energystoragereport. info/abengoa-solar-thermal-energy-storage/.

85.   Martin, Chris & Crawford, Jonathan, "California Power Grid Seen Able to Handle 100% Renewables," *Bloomberg Business*, April 14, 2015, *Michael Picker, president of the California Public Utilities Commission, said the grid already is comfortably managing solar and wind energy that reached as much as 40% of the total a few days last year. In the years ahead, authorities can add the flexibility needed to manage power that flows only when the wind blows or the sun shines. "We could get to 100 percent renewables," Picker said at the Bloomberg New Energy Finance summit in New York on Tuesday. "Getting to 50 percent is not really a challenge."* Available at, http://www.bloomberg.com/news/ articles/2015-04-14/california-power-grid-seen-able-to-handle-100- renewables; *See also,* Hart,E.K., Stoutenburg, E.D. and Jacobson, M.Z., "The Potential of Intermittent Renewables to Meet Electric Power Demand: Current Methods and Emerging Analytical Techniques," *Proceedings of the IEEE*, Vol. 100, No. 2, pp. 322–324, 331, 2012; *See also,* Jacobson, Mark Z., et al., A low-cost solution to the grid reliability problem with 100% penetration of intermittent wind, water,

and solar for all purposes, *Proceedings of the National Academy of Sciences*, publication pending Nov 2015.

86. Akhil, Abbas A., et al., "DOE/EPRI Electricity Storage Handbook in Collaboration with NRECA," *Sandia Report*, SAND2013-5131, July 2013, explores storage technologies in depth. The handbook includes a useful visualization of a portfolio of technologies by providing a schematic demonstrating their potential functions in association with their power and energy relationships. Batteries and flywheels, for example, have a size range from a few kW to 100 MW and have charge/discharge times ranging from seconds to minutes or hours depending on the technology; *See also*, Ferrell, John, "The Next Charge for Distributed Energy," *Institute for Local Self-Reliance (ILSR)*, pp. 17, March 2014.

87. Dragoon, Ken, "Energy Storage Opportunities and Challenges," prepared by *Ecofys* for *EDF Renewable Energy*, April 4, 2014.

88. *Ibid*; *See also*, Dr. Imre Gyuk, program manager of storage research at the U.S. Department of Energy (DOE), "2014 Presentation," keynote speaker at *DOE Energy Storage Workshop* in Oregon, March 19, 2014, Available at, http://www.oregon.gov/energy/pages/energy-stor-age-workshop.aspx

89. Dragoon, Ken, "The State and Promise of Energy Storage," *ElectricityPolicy.com*, April 2014; *See also*, Dragoon, Ken, "Energy Storage Opportunities and Challenges," prepared by *Ecofys* for *EDF Renewable Energy*, April 4, 2014; *See also*, Load Following power plant, *Wikipedia,* https://en.wikipedia.org/wiki/Load_following_power_plant; *See also*, Deign, Jason, "Abengoa: Could molten salt do peaker job?" *Energy Storage Report*, July 1, 2015, Available at, http://ener-gystoragereport.info/abengoa-solar-thermal-energy-storage/.

90. McMahon, Jeff, "Why Tesla batteries are cheap enough to prevent new power plants," *Forbes,* May 5, 2015, Available at, http://www.forbes.com/sites/jeffmcmahon/2015/05/05/why-tesla-batteries-are-cheap-enough-to-prevent-new-power-plants/; *See also*, Naam, Ramez, "Why Energy Storage is About to Get Big—and Cheap, April 14, 2015," Available at, http://rameznaam.com/2015/04/14/

energy-storage-about-to-get-big-and-cheap/; *See also*, Chang, Judy et al., "The value of Distributed Electricity Storage in Texas, Proposed Policy for Enabling Grid-Integrated Storage Investments," *The Brattle Group prepared for ONCOR*, November 2014.

91. Wesoff, Eric, "Large-Scale Energy Storage to Reduce Load in New York City," *GreenTech Media*, February 14, 2014, Available at, http://www. greentechmedia.com/articles/read/Grid-Scale-Energy-Storage-to-Reduce-Load-in-New-York-City; *See also*, John, Jeff St., "From battery cell R&D to megawatt-scale grid testing, NY-BEST's lab is now open for business," *GreenTech Media,* May 12, 2014, Available at, http:// www.greentechmedia.com/articles/read/Testing-Ground-For-New-Energy-Storage-in-New-York; *See also,* "Reforming the energy vision," *New York State,* Available at, http://www3.dps.ny.gov/W/PSCWeb.nsf/ All/26BE8A93967E604785257CC40066B91A; *See also,* "Puerto Rico: AEE introduces storage mandate for renewable energy projects," *Mettering.com*, Dec, 20, 2013, *"Under the new MTR, developers will be required to add a minimum 30% of the installation's contracted capacity in storage as well as the flexibility to keep a minimum 45% of the capacity in reserve for at least one minute, for ramping and frequency control to mitigate the intermittency of renewables generation . . . An AEE statement said the company plans to integrate 600 MW of renewables generation during 2014."* Available at, http://www. metering.com/puerto-rico-aee-introduces-storage-mandate-for-re-newable-energy-projects/; *See also*, Frankel, Dean, "Puerto Rico mandates energy storage for all new renewable development," *SmartGridNews*, January 21, 2014, Available at, http://www.smart-gridnews.com/artman/publish/Technologies_Storage/Puerto-Rico-mandates-energy-storage-for-all-new-renewable-development-6300. html#.U3pzo3mM7fM; *See also,* "Oregon Adopts Energy Storage Mandate," *Nebraskans for Solar,* June 12, 2015, Available at, http:// www.nebraskansforsolar.org/2015/06/12/oregon-adopts-energy-stor-age-mandate/. *See also, "This week Oregon Governor Kate Brown signed the state's first energy storage mandate into law. Assuming the law is authorized by the Oregon Public Utility Commission, electric utilities operating within the state will be required to procure at least 5 mega-watt-hours of energy storage by 2020,"* HB 2193, Available at, https://

olis.leg.state.or.us/liz/2015R1/Downloads/MeasureDocument/HB2193.

92.   Peterman, Hon. Carla J., and Charles, Melicia, "Tapping Into the Potential of Energy Storage," *ElectricityPolicy.com,* March 5, 2014; California Commissioner Carla Peterman and Melicia Charles of the commission explain, *"California is on its way to developing energy storage as one cost-effective solution to support renewables integration, grid reliability, and greenhouse gas reduction;"* California bill AB 215, was passed by the California Public Utilities Commission in October 2013. The law directs the commission to set initial energy storage procurement goals for utilities by 2015 and 2020, respectively. Some of those active in the California grid say they fully expect state policy makers to increase the RPS standard to 50% in the not too distant future.

93.   California Energy Commission Energy Almanac, Available at, http://energyalmanac.ca.gov/electricity/electric_generation_capacity.html.

94.   Weshoff, Eric & John, Jeff St., "Breaking: SCE Announces Winners of Energy Storage Contracts Worth 250 MW," *GreenTech Media*, November 3, 2014, Available at, http://www.greentechmedia.com/articles/read/breaking-sce-announces-winners-of-energy-storage-contracts.

95.   Ferrell, John, "The Next Charge for Distributed Energy," *Institute for Local Self-Reliance* (ILSR), pp. 17, March 2014; Utility managers of Hawaii are learning firsthand the current application and value of energy storage in integrating solar energy onto the grid. They report three main benefits that they are seeing: Storage manages fossil fuels limitations in being able quickly ramp generation up or down when clouds pass by, it provides backup for cloudy days (termed "firming"), and it helps maintain frequency. They show us that the value of storage is more than hypothetical, but is indeed a reality; *See also, Ecofys*; See also DOE/EPRI Handbook.

96.   AES Energy Storage Announces 260 MW of Interconnected Global Projects in Construction or Late Stage Development, The 64 MW battery is rated in terms of flexible load, Available at, http://www.aesenergystorage.com/2015/04/27/

aes-energy-storage-announces-260-mw-interconnected-global-proj-
ects-construction-late-stage-development/; *See also*, Duke Energy's
battery storage projects in Texas wins North American award, Duke
Energy has installed a 36 MW/24 MW advanced lead acid battery
to manage wind farm storage with frequency regulation and New
York has a 20 MW flywheel system in place for frequency regulation,
Available at, http://www.duke-energy.com/news/releases/2013091601.
asp; *See also*, DOE Global Energy Storage Database, Available at,
http://www.energystorageexchange.org.

97.   "Hawaii lawmakers mandate 100 percent renewable energy by 2014,"
        Power Engineering, May 7, 2015, Available at, http://www.power-eng.
        com/articles/2015/05/hawaii-lawmakers-mandate-100-percent-re-
        newable-energy-by-2045.html; *See also*, Trabish, Herman, K., "Hawaii
        lawmakers agree on 100% renewables by 2045," *Utility Dive,* April,
        29, 2015, Available at, http://www.utilitydive.com/news/hawaii-law-
        makers-agree-on-100-renewables-by-2045-mandate/392533/; *See also,*
        Request for Proposals—Energy Storage System, "*To meet its goal of
        adding more renewable generation to the O'ahu grid, Hawaiian Electric
        Company is seeking proposals for one or more large-scale energy
        storage systems able to store 60 to 200 megawatts for up to 30 minutes*",
        Available at, http://goo.gl/o2Wqw9; *See also*, John, Jeff, "Hawaii wants
        200 MW of energy storage for soar, wind grid challenges," *GreenTech
        Media*, May 5, 2014, Available at, http://www.greentechmedia.com/
        articles/read/hawaii-wants-200mw-of-energy-storage-for-solar-wind-
        grid-challenges.

98.   Wesoff, Eric, "Large-scale energy storage to reduce load in New York
        city, with help from SGIP-like incentive program," *GreenTech Media,*
        February 10, 2014, Available at, http://www.greentechmedia.com/
        articles/read/Grid-Scale-Energy-Storage-to-Reduce-Load-in-New-York-
        City; *See also*, Jeff, John St., "Testing Ground for New Energy Storage
        in New York," *GreenTech Media*, Available at, http://www.greentech-
        media.com/articles/read/Testing-Ground-For-New-Energy-Storage-in-
        New-York; This initiative intends transform the electricity infrastruc-
        ture to be less centrally dependent and promote energy efficiency,
        advanced energy management, deeper penetration of renewable

energy and distributed resources such as microgrids, on-site power supplies, and storage; *See also*, New York State, "Reforming the energy vision," May 12, 2014, Available at, http://www3.dps.ny.gov/W/PSCWeb.nsf/All/26BE8A93967E604785257CC40066B91A.

99. Dragoon, Ken, "Energy Storage Opportunities and Challenges," prepared by *Ecofys* for *EDF Renewable Energy*, April 4, 2014, *"The complex interplay between the power system, market structures, and generally low wholesale market prices has complicated the fair valuation of energy storage. Historically, the economics of electric energy storage that justified the construction of the nation's fleet of pumped storage facilities hinged on the wholesale electric market price spread between nighttime and daytime. Value was determined on the basis of the cost of the energy purchased for storage at night and the value received from selling the energy back during the day. However, with low-priced natural gas and an influx of zero (or negative) variable cost renewable generation wholesale market prices are no longer high enough to make an economic case for energy storage based on diurnal price spreads. It is expected that this value will rise again as variable renewable resource penetration increases and zero or negative market price ("oversupply") events increase along with it."* Additionally, because of multiple value streams, such as acting as generation and load, storage can be particularly difficult to value. The electricity system structure is changing such as including demand response result in its true value not being recognized. A more complex valuation is needed.

100. Kempener, Ruud and de Vivero, "Gustavo, Renewables and Electricity Storage, A technology roadmap for REmap 2030," *IRENA Report*, June 2015; *See also*, Carbajales-Dale, M., et al., "Can we afford storage? A dynamic net energy analysis of renewable electricity generation supported by energy storage," *Energy Environ. Sci.*, February 5, 2014, 7, 1538–1544; *See also*, Kaun, B. and Chen, S., "Cost-Effectiveness of Energy Storage in California, Application of the EPRI Energy Storage Valuation Tool to Inform the California Public Utility Commission Proceeding R. 10-12-007," *EPRI, Technical Update*, June 2013.

101. McMahon, Jeff, "Why Tesla batteries are cheap enough to prevent

new power plants," *Forbes,* May 5, 2015, Available at, http://www.
forbes.com/sites/jeffmcmahon/2015/05/05/why-tesla-batteries-are-
cheap-enough-to-prevent-new-power-plants/; *See also*, Naam, Ramez,
"Why Energy Storage is About to Get Big—and Cheap," April 14,
2015, available at, http://rameznaam.com/2015/04/14/energy-stor-
age-about-to-get-big-and-cheap/; *See also*, Chang, Judy et al., "The
value of Distributed Electricity Storage in Texas, Proposed Policy for
Enabling Grid-Integrated Storage Investments," *The Brattle Group
prepared for ONCOR,* November 2014; *See also*, Kempener, Ruud and
de Vivero, Gustavo, "Renewables and Electricity Storage, A technology roadmap for REmap 2030," *IRENA Report,* June 2015.

102. Hales, Roy L., "Solar & Wind = 53% of new US capacity in 2014,"
CleanTechnica, February 3, 2015, Available at, http://cleantechnica.
com/2015/02/03/solar-wind-53-new-us-electricity-capacity-2014/;
See also, Meador, Ron, "Wind and solar projects will power half US
utilities new output 2015," *MinnPost, Earth Journal*, March 13, 2015,
Available at, https://www.minnpost.com/earth-journal/2015/03/wind-
and-solar-projects-will-power-half-us-utilities-new-output-2015 &
Navigant Research, Energy storage Tracker 1Q15, Available at, http://
www.navigantresearch.com/research/energy-storage-tracker-1Q15.

Chapter 4: Efficiency: The Low-Cost Partner To Renewables
103. Text of the Energy Efficiency Improvement Act of 2015, Govtracks.
us, Available at, http://www.govtrack.us/congress/bills/114/s535/text.

104. U.S./California: History of the "all electric home," Available at, http://
www.greenspun.com/bboard/q-and-a-fetch-msg.tcl?msg_id=0063xf;
The campaign was very successful with nearly a million homes
receiving the medallion given the homeowner whose home had an
electric stove, water heater, dishwasher, washing machine and heat.

105. Portman, Shaheen, "Targeted Energy Efficiency Bill Passes Senate
Bipartisan Legislation will save energy, protect the environment,
save consumers money, create jobs," Press Release, March 27, 2015,
Available at, http://www.portman.senate.gov/public/index.cfm/2015/3/
portman-shaheen-targeted-energy-efficiency-bill-passes-senate.

106. Jacobson M.Z. & Delucchi M.A, et al., "100% clean and renewable wind, water, and sunlight (WWS) all-sector energy roadmaps for the 50 United States," *Energy and Environmental Science*, May 2015. *Conversion would reduce each state's end-use power demand by a mean of ~39.3% with ~82.4% of this due to the efficiency of electrification and the rest due to end-use energy efficiency improvements.*

107. "Climate Change in the United States: Benefits of Global Action. United States Environmental Protection Agency, Office of Atmospheric Programs," *EPA 430-R-15-001*, 2015, Available at, http://www2.epa.gov/sites/production/files/2015-06/documents/cirareport.pdf.

108. "National Highway Traffic Safety Administration (NHTSA)," Available at, http://www.nhtsa.gov/cars/rules/rulings/CAFE/alternativefuels/background.htm.

109. Office of the Press Secretary, President Obama Announces Historic 54.5 mpg Fuel Efficiency Standard, The White House, Available at, https://www.whitehouse.gov/the-press-office/2011/07/29/president-obama-announces-historic-545-mpg-fuel-efficiency-standard.

110. Federal Trade Commission, "FTC Seeks Public Input on Proposed Changes to Appliance Labeling RuleEnergy Efficiency Labels Help Consumers Comparison Shop," Available at, https://www.ftc.gov/news-events/press-releases/2012/02/ftc-seeks-public-input-proposed-changes-appliance-labeling-rule.

111. Energy Star, "Special Offers and Rebates from ENERGY STAR Partners," Available at, http://www.energystar.gov/rebate-finder.

112. Green Building Advisor, "Are Energy Efficient Appliances Worth it?," Available at, http://www.greenbuildingadvisor.com/blogs/dept/green-communities/are-energy-efficient-appliances-worth-it.

113. TVA, eScore, "Home Improvement Offer, Energyright solutions," Available at, http://www.energyright.com/residential/escore.html. The program Dave began, now termed eScore, remains in place today.

114. California Energy Commission, Available at, http://www.energy.

ca.gov/efficiency/financing.html; *See also*, Energy Trust of Oregon, Available at, http://energytrust.org/residential/incentives/.

115.   California Energy Commission, Energy Almanac, Available at, http://www.energyalmanac.ca.gov/electricity/total_system_power.html.

Chapter 5: The 21st Century Electric Utility

116.   New York State Energy Planning Board, "Shaping the Future of Energy," *New York State Energy Plan Volume*, 2014 Draft; *See also*, New York State, "Reforming the Energy Vision," Available at, http://www.ny.gov/programs/reforming-energy-vision & http://www3.dps.ny.gov/W/PSCWeb.nsf/All/26BE8A93967E604785257CC40066B91A.

117.   Jacobson and Delucchi, "100% all-sector energy roadmaps for the 50 United States," 2015. *"Year 2050 end-use U.S. all-purpose load would be met with ~30.9% onshore wind, B19.1% offshore wind, ~30.7% utility-scale photovoltaics (PV), ~7.2% rooftop PV, ~7.3% concentrated solar power (CSP) with storage, ~1.25% geothermal power, ~0.37% wave power, ~0.14% tidal power, and ~3.01% hydroelectric power. Based on a parallel grid integration study, an additional 4.4% and 7.2% of power beyond that needed for annual loads would be supplied by CSP with storage and solar thermal for heat, respectively, for peaking and grid stability."*

118.   Bade, Gavin, "6 thought leaders on the future of utility business models & regulation," *Utility Dive*, January 28, 2015, Available at, http://www.utilitydive.com/news/6-thought-leaders-on-the-future-of-utility-business-models-regulation/357635/; *See also*, "2015 State of the Electricity Utility Survey results," *Utility Dive*, January 2015.

119.   Knight, Mark, Sloan, Tom, Zichella, Carl, "Tipping point for transactive energy: the evolving policy and technical challenges," *ElectricityPolicy.com*, 2015, Available at, http://www.electricity-policy.com/Articles/tipping-point-for-transactive-energy-the-evolving-industrys-policy-and-technical-challenges; *See also*, Parks, Leah Y., "An interview with infrastructure guru, farmer, and Kansas State Representative Tom Sloan," *ElectricityPolicy.com*, 2015, Available at, http://www.electricitypolicy.com/Articles/

an-interview-with-infrastructure-guru-farmer-and-kansas-state-representative-tom-sloan.

120. Walton, Robert, "How utilities think they will make their money in the future," *Utility Dive*, February 9, 2015, Available at, http://www.utilitydive.com/news/how-utilities-think-they-will-make-their-money-in-the-future/358845/.

121. Warrick, Joby, "Utilities wage campaign against rooftop solar," *The Washington Post*, March 7, 2015, Available at, http://www.washingtonpost.com/national/health-science/utilities-sensing-threat-put-squeeze-on-booming-solar-roof-industry/2015/03/07/2d916f88-c1c9-11e4-ad5c-3b8ce89f1b89_story.html.

122. http://www.reddykilowatt.org; *See also*, https://en.wikipedia.org/wiki/Reddy_Kilowatt

123. http://www.smecc.org/live_better_electrically_medallion_home.htm.

124. *Ibid.*

125. *Ibid.*

126. Hallock, Lindsey & Rob, Sargent, "Shining Rewards, The Value of Rooftop Solar Power for Consumers and Society," *Environment America Research & Policy Center & Frontier Group*, Summer 2015, Available at, http://www.environmentamerica.org/sites/environment/files/reports/EA_shiningrewards_print.pdf.

127. McKibben, Bill, "Power to the People, Why the rise of green energy makes utility companies nervous," *The New Yorker*, June 29, 2015, Available at, http://www.newyorker.com/magazine/2015/06/29/power-to-the-people.

Chapter 6: Renewable Energy Costs Less, Not More

128. Executive Office of the President of the United States, "The Cost of Delaying Action to Stem Climate Change," *White House Report*, July 2014, Available at, https://www.whitehouse.gov/sites/default/files/docs/the_cost_of_delaying_action_to_stem_climate_change.pdf.

129. Tamminen, Terry, *Lives Per Gallon, The true cost of our oil addiction,* Shearwater Books, Island Press, 2009.

130. Reed, Stanley, "OPEC, Keeping Quotas Intact, Adjusts to Oils New Normal," *New York Times,* June 5, 2015, Available at, http://www.nytimes.com/2015/06/06/business/international/opec-oil-prices.html?_r=0.

131. "US shale boom may be over by the end of 2015-OPEC," March 17, 2015, Available at, http://rt.com/business/241377-us-shale-boom-end/; *See also,* Cassidy, John, "OPEC's Warf on Fracking is Good News for the Rest of US," *The New Yorker,* Dec 1, 2014, available at, http://www.newyorker.com/news/john-cassidy/opecs-holiday-present-helps-every-one-except-frackers; *See also,* Phillips, Mathew, "The American Oil Boom Won't Last Long at $65 Per Barrel," *Bloomberg Business,* Dec, 01, 2014, Available at, http://www.bloomberg.com/bw/articles/2014-12-01/can-the-us-fracking-boom-survive-with-oil-65-per-barrel; *See also,* Crow, Alexis, "Sheik vs. Shale: the future of OPEC," *The World Post, Huffington Post,* Jan 01, 2015, Available at, http://www.huffingtonpost.com/alexis-crow/shale-opec_b_6414912.html.

132. EIA, "U.S. Natural Gas Citygate Price," *EIA,* Available at, http://www.eia.gov/dnav/ng/hist/n3050us3m.htm; *See also,* Earth Policy Institute, Available at, http://www.earth-policy.org.

133. Horn, Steve, "The uncertain future of Shale Gas Casts Doubt on US Hydraulic Fracking Production Numbers," *Global Research,* Oct 31, 2014, Available at, http://www.globalresearch.ca/the-uncertain-future-of-shale-gas-report-casts-doubt-on-us-hydraulic-fracking-production-numbers/5410981; *See also,* Hughes, David J, "Drilling, "Deeper: A Reality Check on U.S. Government Forecasts for a Lasting Tight Oil & Shale Gas Boom," *Post Carbon Institute,* October 2014.

134. IRENA, "Renewable Power Generation Costs," *IRENA Report,* January 2015, Available at, http://www.irena.org/DocumentDownloads/Publications/IRENA_RE_Power_Costs_2014_report.pdf; *See also,* Elliston, Ben, MacGill Iain, & Diesendorf Mark, et al., "Comparing least cost scenarios for 100% renewable electricity with low emission fossil fuel scenarios in the Australian National Electricity Market,"

Renewable Energy, Vol. 66, pp. 196-204, 2014.

135.   *Ibid.*

136.   USGS, "Advantages of Hydroelectric Power Production and Usage,"
Available at, http://water.usgs.gov/edu/hydroadvantages.html; *See
also,* Turner, GM, Elliston, B, Diesendorf, M, "Impacts on the biophysical economy and environment of a transition to 100% renewable
electricity in Australia," *Energy Policy,* Vol. 54, pp. 288–299, 2013.

137.   IRENA, "Renewable Power Generation Costs in 2014," *IRENA Report,*
January 2015. Available at, www.irena.org/publications; *See also,*
Cardwell, Dianne, "Solar and Wind Energy Start to Win on Price
vs. Conventional Fuels," *New York Times,* Nov 23 2014, Available
at, http://www.nytimes.com/2014/11/24/business/energy-environ-
ment/solar-and-wind-energy-start-to-win-on-price-vs-conventional-
fuels.html?_r=0; *See also,* Lazard, "Lazard Levelized Cost of Energy
Analysis-Version 8.0," September 2014; *See also,* Jacobson M.Z. &
Delucchi M.A, et al., "100% clean and renewable wind, water, and
sunlight (WWS) all-sector energy roadmaps for the 50 United States,"
Energy and Environmental Science, May 2015.

138.   Bronski, Peter, Lilienthal ,Peter, and Crowdis, Mark, et al., "The
Economics of Grid Defection When and Where distributed Soar
Generation Plus Storage Competes with Traditional Utility Service,"
Rocky Mountain Institute, February 2014, Available at http://www.
rmi.org; *See also,* Wile, Rob, "GOLDMAN: Solar Is On The Way To
Dominating The Electricity Market, And The World Has Elon Musk
To Thank," *Business Insider,* March 18, 2014. Available at, http://www.
businessinsider.com/goldman-on-solar-and-elon-musk-2014-3; *See
also,* Romm, Joe, "Must-See Chart: Cost Of PV Cells Has Dropped
An Amazing 99% Since 1977, Bringing Solar Power To Grid Parity,"
Think Progress, October 6, 2013, Available at, http://thinkprogress.org/
climate/2013/10/06/2717791/cost-pv-cells-solar-power-grid-parity/; *See
also,* Romm Joe, "Solar Power Is Now Just As Cheap As Conventional
Electricity In Italy And Germany," *Think Progress,* March 24, 2014,
Available at, http://thinkprogress.org/climate/2014/03/24/3418145/
solar-grid-parity-italy-germany/.

139. IRENA, "Renewable Power Generation Costs in 2014," *International Renewable Energy Agency (IRENA) Report*, January 2015. Available at, http://www.irena.org/publications; *See also*, Cardwell, Dianne, "Solar and Wind Energy Start to Win on Price vs. Conventional Fuels," *New York Times*, Nov 23 2014, Available at, http://www.nytimes.com/2014/11/24/business/energy-environment/solar-and-wind-energy-start-to-win-on-price-vs-conventional-fuels.html?_r=0; *See also*, Lazard, "Lazard Levelized Cost of Energy Analysis-Version 8.0," September 2014; Jacobson M.Z. & Delucchi M.A, et al., "100% clean and renewable wind, water, and sunlight (WWS) all-sector energy roadmaps for the 50 United States," *Energy and Environmental Science*, May 2015.

140. IRENA, "Renewable Power Generation Costs in 2014," *IRENA Report*, January 2015; *See also*, Jacobson and Delucchi et al., "100% all-sector energy roadmaps for the 50 United States," 2015; *See also*, Elliston, Ben, MacGill Iain, & Diesendorf Mark, et al., "Comparing least cost scenarios for 100% renewable electricity," *Renewable Energy,* 2014; *See also*, Turner, GM, Elliston, B, Diesendorf, M, "Impacts on the biophysical economy and environment of a transition to 100% renewable electricity in Australia," *Energy Policy*, Vol. 54, pp. 288-299, December 2012.

141. Jacobson and Delucchi et al, "100% all-sector energy roadmaps for the 50 United States," 2015. Costs due to climate change include coastal flood and real estate damage costs, energy-sector costs, health costs due to heat stress and heat stroke, influenza and malaria costs, famine costs, ocean acidification costs, increased drought and wildfire costs, severe weather costs, and increased air pollution health costs (costs are in 2013 dollars).

142. Cuff, Madeleine, "The Top 7 reasons businesses should fear climate change," *GreenBiz*, August 10, 2015, Available at, http://www.greenbiz.com/article/top-7-reasons-businesses-should-fear-climate-change

143. Executive Office of the President of the United States, "The Cost of Delaying Action to Stem Climate Change," *White House Report*, July 2014; *See also*, Tamminen, Terry, *Lives Per Gallon*, 2009; *See also*,

Jacobson and Delucchi, "100% all-sector energy roadmaps for the 50 United States," 2015.

144. Barney Jeffries, Yvonne Deng, et al., "The Energy Report-100% Renewable Energy by 2050," *WWF, ECOFYS & OMA Technical Report* (2011), The joint study by WWF, ECOFYS and OMA concluded that it was possible to supply the world's energy needs with 100% renewable by 2050; *See also,* IRENA, "International Renewable Energy Agency," Available at, http://www.irena.org/home/index.aspx; *See also,* Elliston, Ben, MacGill Iain, & Diesendorf Mark, et al., "Comparing least cost scenarios for 100% renewable electricity," 2014; *See also,* Turner, GM, Elliston, B, Diesendorf, M, "Impacts on the biophysical economy and environment of a transition to 100% renewable electricity in Australia," 2012; *See also,* Jacobson and Delucchi et al., "100% all-sector energy roadmaps for the 50 United States," 2015.

145. Smialek, Jeanna, "These Will Be the World's 20 Largest Economies in 2030," *Bloomberg Business,* April 10, 2015, Available at, http://www.bloomberg.com/news/articles/2015-04-10/the-world-s-20-largest-economies-in-2030.

146. Volcovici, Valerie and Brunnstrom, David, "China Puts $6 trillion price tag on its climate plan," *Reuters,* June 23, 2015, Available at, http://mobile.reuters.com/article/idUSL1N0Z92A920150623?irpc=932.

147. Jacobson and Delucchi et al, "100% all-sector energy roadmaps for the 50 United States," 2015. *"The total up-front capital cost of the 2050 WWS system is ~$13.4 trillion (~$2.08 mil. per MW);" See also,* GDP (current US dollars), The World Bank, Available at, http://data.worldbank.org/indicator/NY.GDP.MKTP.CD; *See also,* "US defense spending history," *USGovermenteSpending.com,* available at, http://www.usgovernmentspending.com/defense_spending.

148. Jacobson and Delucchi et al, "100% all-sector energy roadmaps for the 50 United States," 2015. *"Conversion would reduce each state's end-use power demand by a mean of ~39.3% with ~82.4% of this due to the efficiency of electrification and the rest due to end-use energy efficiency improvements . . . Over all 50 states, converting would provide ~3.9 million 40-year construction jobs and ~2.0 million 40-year operation*

jobs for the energy facilities alone, the sum of which would outweigh the ~3.9 million jobs lost in the conventional energy sector. Converting would also eliminate ~62,000 (19,000–115,000) U.S. air pollution premature mortalities per year today and ~46,000 (12,000–104,000) in 2050, avoiding ~$600 ($85–$2400) bil. per year (2013 dollars) in 2050, equivalent to ~3.6 (0.5–14.3) percent of the 2014 U.S. gross domestic product. Converting would further eliminate ~$3.3 (1.9–7.1) tril. per year in 2050 global warming costs to the world due to U.S. emissions.). The sum of all states' emissions cause a net positive damage to the U.S. as a whole (with total damage caused by all states' emissions in 2050 of $265 bil. per year in 2013 dollars). These plans will result in each person in the U.S. in 2050 saving ~$260 (190–320) per year in energy costs ($2013 dollars) and U.S. health and global climate costs per person decreasing by ~$1500 (210–6000) per year and ~$8300 (4700–17 600) per year, respectively ... The total up-front capital cost of the 2050 WWS system is ~$13.4 trillion (~$2.08 mil. per MW)."

PART II: OBSTACLES

Chapter 7: The Hidden Truth about Natural Gas

149. Lawrence Livermore National Laboratory, "Estimated US Energy use in 2014: 97.4 Quads," available at: https://flowcharts.llnl.gov/commodities/carbon.

150. *Ibid.*

151. *Ibid.*

152. *Ibid. See also,* https://www2.ucar.edu/atmosnews/news/5292/switching-coal-natural-gas-would-do-little-global-climate-study-indicates

153. We explain why natural gas produces enough carbon dioxide to cause us to blow past our carbon budget and why we must achieve zero emissions by 2050 in more detail in the chapter 8, "Leave it in the Ground."

154. IPCC, 2013, *Climate Change 2013: The Physical Science Basis.*

Contribution of Working Group I to the Fifth Assessment Report of the Intergovernmental Panel on Climate Change [Stocker, T.F., D. Qin, G.-K. Plattner, M. Tignor, S.K. Allen, J. Boschung, A. Nauels, Y. Xia, V. Bex and P.M. Midgley (eds.)]. Cambridge University Press, Cambridge, United Kingdom and New York, NY, USA, Hereinafter referred to in this chapter, IPCC, WGI, AR5 2013; *See also*, EPA, "Inventory of U.S. Greenhouse Gas Emissions and Sinks: 1990-2012," *U.S. Environmental Protection Agency Report,* April 15, 2015, Often referred to in this chapter as, EPA 2015 Inventory; *See also*, Busch, Chris & Gimon, "Eric, Natural Gas versus Coal: Is Natural Gas Better for the Climate?" *The Electricity Journal*, Vol. 27, Issue 7, pp. 1040–6190, Aug/Sept 2014; *See also*, Shindell, Drew T. et al., "Improved Attribution of Climate Forcing to Emissions," *Science*, Vol. 326, pp. 716, October 2009. The factor, 84 is called the global warming potential (GWP) and compares methane to carbon dioxide on a pound per pound basis. The IPCC, WGI, AR5 2013 determined a GWP for methane compared to CO_2 to be 84 over a 20-year period; In the EPA 2015 Inventory, the global warming potential of 25 over 100 years is used for its estimates of natural gas's CO_2eq emissions. However recent analysis indicates it is necessary to consider the GWP over a 20-year period, which has been estimated to be between 84–105 times of that of carbon dioxide.

155. IPCC, WGI, AR5 2013; *See also*, Schindell, Drew et al., "Simultaneously Mitigating Near-Term Climate Change and Improving Human Health and Food Security," *Science*, Vol. 335, pp. 183–189, January 13, 2012; *See also*, UNEP &WMO, "Integrated Assessment of Black Carbon and Tropospheric Ozone," *United Nations Environmental Program & World Meteorological Organization,* UNON/Publishing Services Section/Nairobi, 2011; *See also*, Shaefer, K., et al., "Amount and timing of permafrost carbon release in response to climate warming," *Tellus* B, Vol. 63, Issue 2, pages 165–180, April 2011; *See also*, Schuur E. A. G., et al., "Vulnerability of Permafrost Carbon to Climate Change: Implications for the Global Carbon Cycle," *BioSci.*, Vol. 58, No.8, pp. 701–714, September 2008; (Many studies that call for an immediate end to fuels that produce methane to secure our climate future).

156. Shoemaker et. al, "What Role for Short-Lived Climate Pollutants in

Mitigation Policy?" *Science*, Vol. 342, pp. 1323-1324, December 13, 2013; *See also*, Busch, Chris & Gimon, Eric, "Natural Gas versus Coal," *The Electricity Journal*, Aug/Sept 2014, *"The impacts of climate change in the near term are also of interest to policymakers. The impacts of climate change are being experienced today. Wildfires are intensifying; sea level rise is accelerating; glaciers are melting. As ice melts, the reflectivity of the earth's surface is decreased. This leads to more heat being absorbed and less reflecting back and escaping the earth's atmosphere, in what is known as a positive feedback loop. Reducing short-lived GHGs such as methane decreases the current and near-term damages from global warming. This is valuable in its own right, and creates time for greater technological innovation to occur and to prepare for necessary adaptations."* E.g., some of the most disastrous impacts are; sea level rise, glacial melting, and intensity of wildfires as well as the potential and extent of positive feedback loops and tipping points associated with these environmental impacts.

157. To determine the percentage values were obtained from EPA 2015 Inventory Report. (Note: the amount of methane emitted from natural gas systems in EPA calculations assumes a conservative 1.5% leakage rate estimate). To determine that methane accounts for 26% of greenhouse gases we used the EPA natural gas system estimate, which assumed a GWP of 25 at a 100-year time frame, and adjusted that to an 84 GWP at a 20-year timeframe. This results in total methane CO_2eq emissions at 2138 CO_2eq and total US emissions at 8175 CO_2eq. Therefore, the methane CO_2eq emissions, over a 20-year period, cause 26% of the total problem (2138/8175=26%). Also note: The percent of CO_2eq from methane may actually be greater than 26%. It is estimated that total natural gas emissions have been underestimated as we will describe in greater detail in the next section of this chapter.

158. Larsen, Kate, Delgado, Michael, and Marsters, Peter, "Untapped Potential: Reducing Global Methane Emissions from Oil and Natural Gas Systems," *Rhodium Group Report*, April 2015, Available at: http://www.edf.org/climate/rhodium-group-report-global-oil-gas-methane-emissions?_ga=1.219423705.1787517823.142785108

8.

159. "Bottom up" studies identify finer details, but given the complexity of the natural gas system, large extrapolations are put into question. "Top down" studies give a better overall picture (snapshot), but they also measure other methane sources that must be removed from the results and there is variation from field to field.

160. Larsen, Kate, et al., "Untapped Potential," *Rhodium Group Report*, April 2015; *See also*, Busch, Chris & Gimon, Eric, "Natural Gas versus Coal," *The Electricity Journal*, Aug/Sept 2014; *See also*, Howarth, Robert, Shindell, Drew, et al., "Methane Emissions from Natural Gas Systems Background Paper Prepared for the National Climate Assessment," Reference number 2011–0003, February 25, 2012; *See also*, Howarth R.W., "A bridge to nowhere: methane emissions and the greenhouse gas footprint of natural gas," *Energy Science & Engineering*, Vol. 2 no. 2, pp. 47–60, June 2014; *See also*, Allen, David T., et al., "Measurements of methane emissions at natural gas production sites in the United States," *PNAS*, Vol. 110 no. 44, pp. 17768–17773, October 29, 2013.

161. Miller, Scot M., et al., "Anthropogenic emissions of methane in the United States," *PNAS*, Vol. 110 no. 50, pp. 20018–20022, December 10, 2013; *See also*, Petrón, G., et al., "Hydrocarbon emissions characterization in the Colorado Front Range: A pilot study," *Journal of Geophysical Research: Atmospheres*, Vol. 117 no.1, pp. 236–242, January 16, 2013; *See also*, Peischl, J., et al., "Quantifying sources of methane using light alkanes in the Los Angeles basin, California," *Journal of Geophysical Research: Atmospheres*, Vol. 118 no. 10, pp. 4974–4990, May 27, 2013; *See also*, Karion, A., et al., "Methane emissions estimate from airborne measurements over a western United States natural gas field," *Geophysical Research Letters*. Vol. 40 no. 16, pp. 4393–4397, August 28, 2013; *See also*, Caulton, Dana R., et al., "Toward a better understanding and quantification of methane emissions from shale gas development," *PNAS*, Vol. 111 no. 17, April 29, 2014; *See also*, Peischl, J., et al., "Quantifying atmospheric methane emissions from the Haynesville, Fayetteville, and northeastern Marcellus shale gas production regions," *Journal of Geophysical Research: Atmospheres*,

Vol. 120 no. 5, pp. 2119–2139, March 16, 2015; *See also*, Schneising, Oliver, et al., "Remote sensing of fugitive methane emissions from oil and gas production in North American tight geologic formations," *Earth's Future*, Vol. 2, pp. 548–558, October 6, 2014; *See also*, K. J. Wecht, et al., "Spatially resolving methane emissions in California: constraints from the CalNex aircraft campaign and from present (GOSAT, TES) and future (TROPOMI, geostationary) satellite observations," Atmos. *Chem. Phys.*, Vol. 14, pp. 8173–8184, 2014.

162.   Bruckner T., et. al, *Energy Systems. In: Climate Change 2014: "Mitigation of Climate Change. Contribution of Working Group III to the Fifth Assessment Report of the Intergovernmental Panel on Climate Change [Edenhofer, O., et. al]*. Cambridge University Press, Cambridge, United Kingdom and New York, NY, USA, 2014, Available at, http://www.ipcc.ch/pdf/assessment-report/ar5/wg3/ipcc_wg3_ar5_chapter7.pdf; *See also*, Larsen, Kate, et al., "Untapped Potential," *Rhodium Group Report*, April 2015; *See also*, Busch, Chris & Gimon, Eric, "Natural Gas versus Coal," *The Electricity Journal*, Aug/Sept 2014; *See also*, Schweitzke S, Griffin WM, Matthews HS, and Bruhwiler LMP, "Natural gas fugitive emissions rates constrained by global atmospheric methane and methane," *Environmental Science and Technology*, Vol. 48 no.14, pp. 7714–7722, July 2014; *See also*, Brandt, A. R., et al., "Methane leaks from north American gas systems," *Science*, Vol. 343, p. 733–735., February 14, 2014.

163.   Brandt, A. R., et al., "Methane leaks from north American gas systems," February 14, 2014; *See also*, Turner, A. J., et al., "Estimating global and North American methane emissions with high spatial resolution using GOSAT satellite data," *Atmos. Chem. Phys. Discuss*, Vol. 15, pp. 4495–4536, February 18, 2015; *See also*, Romm, Joe, "Bridge Out: Bombshell Study Finds Methane Emissions From Natural Gas Production Far Higher Than EPA Estimates," *Climate Progress*, November 25, 2013, Available at, http://thinkprogress.org/climate/2013/11/25/2988801/study-methane-emissions-natural-gas-production/; *See also*, "PSE Healthy Energy Methane Emissions from Modern Natural gas Development," *PSE Healthy Energy Science Summary*, March 2012, Available at, http://www.psehealthyenergy.org/

data/Methane_Science_Summary.pdf. *See also*, http://www.source-watch.org/index.php?title=Natural_gas_transmission_leakage_rates

164. Marchese, Anthony J., et al., "Methane Emissions from United States Natural Gas Gathering and Processing," *Env. Sci and Technol*, August 18, 2015, Available at, http://pubs.acs.org/doi/abs/10.1021/acs.est.5b02275

165. EPA, "US EPA, Inventory of U.S. Greenhouse Gas Emissions and Sinks: 1990–2013," *US EPA*, pp. ES–12, April 2015.

166. Markey, Ed, "Markey Report: Leaky Natural Gas Pipelines Costing Consumers Billions," August 1, 2013, Available at, http://www.markey.senate.gov/news/press-releases/markey-report-leaky-natural-gas-pipelines-costing-consumers-billions; *See also*, Phillips, N.G., et al., "Mapping urban pipeline leaks: Methane leaks across Boston," *Environmental Pollution (2012)*, Available at, http://dx.doi.org/10.1016/j.envpol.2012.11.003; *See also*, https://news.wsu.edu/2013/04/10/natural-gas-methane-emissions-focus-of-new-study/#.VIEw1IeM6-I); *See also*, Jessie Coleman, et al., "Methane Emissions from the Oil and Gas Industry: Under-reported and Under-regulated," *Greenpeace Report*, March 2015. *See also*, "List of pipeline accidents," *Wikipedia*, Available at, http://en.wikipedia.org/wiki/List_of_pipeline_accidents.

167. Ingraffa & Howarth, "Methane Emissions from Natural gas System," February 25, 2012; *See also*, Howarth R.W., A bridge to nowhere: methane emissions and the greenhouse gas footprint of natural gas, June 2014.

168. McKain, Kathryn, et al., "Methane emissions from natural gas infrastructure and use in the urban region of Boston, Massachusetts," *PNAS*, vol. 112 no. 7, 1941–1946, February 17, 2015. The study looked at leakage in the downstream components of the natural gas system, including transmission, distribution, and end-use. "... *loss rate to the atmosphere from all downstream components of the natural gas system, including transmission, distribution, and end use, was 2.7 ± 0.6% in the Boston urban region, with little seasonal variability. This fraction is notably higher than the 1.1% implied by the most closely comparable emission inventory;*" *See also*, Phillips, Nathan G., et

al., "Mapping urban pipeline leaks: Methane leaks across Boston," *Environmental Pollution,* Vol. 173, pp. 1–4, 2013; *See also,* Lamb, Brian K., "Direct Measurements Show Decreasing Methane Emissions from Natural Gas Local Distribution Systems in the United States," *Environmental Science and Technology,* Vol. 49 no. 8, pp 5161–5169, March 31, 2015. The study finds that leaks in the distribution system are less than 20 years ago, but do vary by region and age of the city. They report that, *"the upper confidence limit accounts for the skewed distribution of measurements, where a few large emitters accounted for most of the emissions. This emission estimate is 36% to 70% less than the EPA 2011 Inventory, (based largely on 1990s emission data), and reflects significant upgrades at metering and regulating stations, improvements in leak detection and maintenance activities, as well as potential effects from differences in methodologies between the two studies."*

169. McKain, Kathryn, et al., "Methane emissions from natural gas infrastructure and use in the urban region of Boston, Massachusetts," *PNAS,* Vol. 112, No 7, pp. 1941-1946. February 17, 2015. *"Most recent analyses of the environmental impact of natural gas have focused on production, with very sparse information on emissions from distribution and end use. This study quantifies the full seasonal cycle of methane emissions and the fractional contribution of natural gas for the urbanized region centered on Boston."*

170. EPA, "Inventory of U.S. Greenhouse Gas Emissions and Sinks; 1990-2013," *US EPA,* April 2015; *See also,* "Lawrence Livermore National Laboratory, 2015 Energy and Carbon Flow Charts," Available at, https://flowcharts.llnl.gov; *See also,* http://www.epa.gov/methane/gasstar/basic-information/index.html. We compared total carbon dioxide equivalent emissions (CO_2eq) between natural gas and coal using data from Department of Energy (DOE) Lawrence Livermore flow charts and US EPA 2015 Inventory. We use 2013 data from reporting on CO_2 and methane (CO_2eq) emissions and reported data on total quads of energy used. Coal mining CO_2eq methane emissions were obtained from the US EPA 2015 Inventory. Total coal emissions includes 1720 Million Metric Tons carbon dioxide (MMT

CO_2) for electricity generation and industrial and commercial uses + 64.6 MMT CO_2eq from methane from coal mining. Total natural gas carbon dioxide emissions included in 1390 MMT CO_2 for electricity generation, residential, commercial, industrial and transportation uses. To determine the amount of methane emissions at different leakage rates from natural gas we used the total Quads of natural gas (electrical, residential, commercial, and industrial used) from the DOE flow charts and multiplied by estimated methane leakage rates to obtain the total Quads of leaked natural gas produced at a given leakage rate. We assumed 95% of the natural gas leaked is methane. We then convert Quads of methane produced from the natural gas system to CO_2eq. We assumed a GWP of 84.

171. "The White House, Climate Action Plan Strategy to Reduce Methane Emissions," March 2014, Available at, http://www.whitehouse.gov/sites/default/files/strategy_to_reduce_methane_emissions_2014-03-28_final.pdf; *See also,* https://www2.ucar.edu/atmosnews/news/5292/switching-coal-natural-gas-would-do-little-global-climate-study-indicates

172. EPA, "Oil and Natural Gas Sector: New Source Performance Standards and National Emission Standards for Hazardous Air Pollutants Reviews, Federal Register, Rules and Regulations," *US EPA,* Vol. 77, no. 159, p. 49490, Thursday, August 16, 2012, Available at, http://www.gpo.gov/fdsys/pkg/FR-2012-08-16/pdf/2012-16806.pdf; *See also,* EPA, "Federal Registry, Greenhouse Gas Reporting Rule: 2015 Revisions and Confidentiality Determinations for Petroleum and Natural Gas Systems; Proposed Rule," *US EPA,* Vol. 79, No. 236, December 9, 2014, Available at, http://www.gpo.gov/fdsys/pkg/FR-2014-12-09/pdf/2014-28395.pdf; *See also,* EPA, "Proposed Rule submitted for publication in the Federal Register," *US EPA,* August 18, 2015, *EPA Administrator, Gina McCarthy, signed the following notice on 8/18/2015, and EPA is submitting it for publication in the Federal Register (FR). Refer to the official version in a forthcoming FR publication, which will appear on the Government Printing Office's FDSys website (http://gpo.gov/fdsys/ search/home.action) and on Regulations.gov (http://www.regulations. gov) in Docket No. EPA–HQ–OAR–2010-0505. Once the official version of*

this document is published in the FR, this version will be removed from the Internet and replaced with a link to the official version. Available at, http://www.epa.gov/airquality/oilandgas/pdfs/og_nsps_pr_081815.pdf

173. ICF, "Economic Analysis of Methane Emissions Reduction Opportunities in the US Onshore Oil and Natural Gas Industries," *ICF Report prepared for the EDF,* March 2014. Available at: https://www.edf.org/sites/default/files/methane_cost_curve_report.pdf, *"Almost all of this growth is from the oil sector whereas the net emissions for the gas sector are almost unchanged (Figure 3-1). Growth from new sources in the gas sector is offset by NSPS reductions, and reductions from existing sources such as continuing replacement of cast iron mains and turnover of high-emitting pneumatic devices. Despite the overall growth, nearly 90% of the emissions in 2018 come from existing sources (sources in place as of 2011). This study takes in to account the increased increase efficiency of controlling emissions at the new wells."*

174. Hollinger, J., "An Update on EPA's Approach to Methane Emissions from the Oil & Gas Sector—Including a Summary of the Agency's Proposed New Reporting Rule," *Energy Business Law,* December, 2014; Available at, http://www.energybusinesslaw.com/2014/12/articles/environmental/an-update-on-epas-approach-to-methane-emissions-from-the-oil-gas-sector-including-a-summary-of-the-agencys-pro-posed-new-reporting-rule/?utm_source=Mondaq&utm_medium=syn-dication&utm_campaign=LinkedIn-integration; *See also,* Hollinger, J., "The President's Methane Reduction Strategy—Here's What Energy Companies Need to Know," *Energy Business Law,* April, 2014, Available at, http://www.energybusinesslaw.com/2014/04/articles/natural-gas/the-presidents-methane-reduction-strategy-heres-what-energy-com-panies-need-to-know/, According to Hollinger, the *"(Strategy) that sets forth a multi-pronged plan for reducing methane emissions both domestically and globally. Domestically, the plan is to focus on four sources of methane—the oil and gas sector, coal mines, agriculture and landfills—and to pursue a mix of regulatory actions with respect to those sources... For the oil and gas sector, the Strategy indicates that the federal government will focus primarily on encouraging voluntary efforts to reduce methane emissions... But the Strategy also identifies*

two areas of potential mandatory *requirements... First, a draft rule on minimizing venting and flaring on public lands ... Second ... Environmental Protection Agency (EPA) will decide this fall whether to propose any mandatory methane control requirements on oil and gas production companies."*

175. Wigley, Tom M.L., "Coal to gas: the influence of methane leakage," *Climate Change*, Vol. 108, no. 3, pp. 601–608, October 2011; *See also*, Myhrvold, N.P. and Caldeira K., "Greenhouse gases, climate change and the transition from coal to low-carbon electricity," *Environmental Research Letters*, Vol. 7 no. 1, pp. 014019, February 16, 2012; *See also*, Shearer, Christine, et al., "The effect of natural gas supply on US renewable energy and CO_2 emissions," *Environmental Research Letters*, Vol. 9, 094008, September 24, 2014.

Chapter 8: Keep It In The Ground

176. Skone, Timothy J. & Cooney, Gregory, et al., "Life Cycle Greenhouse Gas Perspective on Exporting Liquefied Natural Gas from the United States," *US DOE National Energy Technology Laboratory (NETL)*, Prepared by Energy Sector Planning and Analysis (ESPA), DOE/NETL-2014/1649, May 29, 2014; *See also*, Camuzeaux J.R. and Alvarez R.A. et al., "Influence of Methane Emissions and Vehicle Efficiency on the Climate Implications of Heavy-Duty Natural Gas Trucks," *Environ. Sci. Technol.*, Vol. 49 no. 11, pp 6402–6410, May19, 2015; *See also*, Alvarez, Ramón A., et al., "Greater focus needed on methane leakage from natural gas infrastructure," *PNAS*, Vol. 109 no. 17, pp. 6435–6440, February 13, 2012; *See also*, Romm, Joe, "Energy Department Bombshell: LNG Has No Climate Benefit For Decades, IF EVER," Climate Progress, June 4, 2014, Available at, http://thinkprogress.org/climate/2014/06/04/3443211/energy-department-lng-no-climate-benefits/.

177. IPCC, 2013: *Climate Change 2013: The Physical Science Basis. Contribution of Working Group I to the Fifth Assessment Report of the Intergovernmental Panel on Climate Change* [Stocker, T.F., D. Qin, G.-K. Plattner, M. Tignor, S.K. Allen, J. Boschung, A. Nauels, Y.

Xia, V. Bex and P.M. Midgley (eds.)]. Cambridge University Press, Cambridge, United Kingdom and New York, NY, USA, pp. 26–28, 103; Hereinafter in this chapter referred to IPCC, WGI, AR5 2013; In this book we consider a greenhouse gas budget, which takes into account greenhouse gases released such as methane and not just carbon dioxide. Only considering a budget based on carbon dioxide does not give an accurate picture of when we will raise the Earth's temperature by 2°C. According to the IPCC report, *"When accounting for the non-CO_2 forcings as in the RCP scenarios, compatible carbon emissions since 1870 are reduced to about 900 PgC, 820 PgC and 790 PgC to limit warming to less than 2°C since the period 1861–1880 with a probability of >33%, >50%, and >66%, respectively . . . (1PgC = 1GtC). Non-CO_2 forcing constituents are important, requiring either assumptions on how CO_2 emission reductions are linked to changes in other forcings, or separate emission budgets and climate modelling for short-lived and long-lived gases. So far, not many studies have considered non-CO_2 forcings. Those that do consider them found significant effects, in particular warming of several tenths of a degree for abrupt reductions in emissions of short-lived species, like aerosols . . . Accounting for an unanticipated release of GHGs from permafrost or methane hydrates, not included in studies assessed here, would also reduce the anthropogenic CO_2 emissions compatible with a given temperature target. Requiring a higher likelihood of temperatures remaining below a given temperature target would further reduce the compatible emissions."*

178. United Nations Framework Convention on Climate Change, "Copenhagen Climate Change Conference," December 2009, Available at, http://unfccc.int/meetings/copenhagen_dec_2009/meeting/6295.php; *See also,* UNFCC, *Report of the Conference of the Parties on its fifteenth session,* held in Copenhagen from 7 to 19 December 2009, Framework Convention on Climate Change, United Nations, March 30, 2010, Available at: http://unfccc.int/resource/docs/2009/cop15/eng/11a01.pdf & http://unfccc.int/meetings/copenhagen_dec_2009/items/5262.php; *"The Conference of the Parties (COP), at its fifteenth session, took note of the Copenhagen Accord of 18 December 2009 by way of decision 2/CP.15."* 114 parties agreed upon the Accord. Since

the report of the COP was issued other parties have come forward "
*... expressing their intention to be listed as agreeing to the Accord ...
As result the total number of Parties that have expressed their intention to be listed as agreeing to the Accord is 141."* The Agreement is
nonbinding. It also did not provide any agreement on how to achieve
the 2°C goal in practical terms. It included considering limiting the
temperature increase to 1.5°C—*"a key demand made by vulnerable
developing countries."*

179.   IPCC, WGI, AR5 2013, pp. 26–28, 103; *See also*, UNFCC, *Report of
the Conference of the Parties on its fifteenth session*, March 30, 2010,
Available at: http://unfccc.int/resource/docs/2009/cop15/eng/11a01.
pdf; *See also*, Anderson, K. and Bows, A., "Beyond 'dangerous'
climate change: emission scenarios for a new world," *Philosophical
Transactions A of the Royal Society*, No. 369, pp. 20–44, 2011; *See also*,
Anderson, Kevin, "Climate change going beyond dangerous—Brutal
numbers and tenuous hope," *Development Dialogue Climate*, What
Next Volume III, Development and Equity, Dag Hammarskjöld
Foundation and the What Next Forum, no. 61, September 2012.

180.   "Paris 2015 UN Climate Change Conference, 21st Session of the
Conference of the Parties to the United Nations Framework
Convention on Climate Change (COP21/CMP11), "Paris 2015" from
November 30th to December 11th," Available at: http://www.cop21.
gouv.fr/en. The Copenhagen accord is subject to a review in 2015
with the intention of developing a legally binding agreement. The
21st session of the Conference of the Parties to the United Nations
Framework Convention on Climate Change (UNFCCC), or COP21,
will be held from November the 30th to December the 11th, 2015 on
the Paris-Le Bourget site *"... The aim is to reach, for the first time,
a universal, legally binding agreement that will enable us to combat
climate change effectively and boost the transition towards resilient,
low-carbon societies and economies. To achieve this, the future agreement must focus equally on mitigation—that is, efforts to reduce greenhouse gas emissions in order to limit global warming to below 2°C."*

181.   IPCC, WGI, AR5 2013, pp. 26–28, 103. "Equivalent CO_2 (CO_2eq) is the
concentration of CO_2 that would cause the same level of (warming)

as a given type and concentration of different greenhouse gases.

182. IPCC, WGI, AR5 2013, pp. 26–28, 103.

183. Nicholas Stern, Hans Joachim Schellnhuber, Jeffrey Sachs et al., "Earth Statement," *Earth League*, 2015, Available at: http://www. the-earth-league.org & http://earthstatement.org & http://www. theguardian.com/environment/2015/apr/22/earth-day-scientists-warning-fossil-fuels-; Our calculations are consistent with The Earth League findings. To determine emission levels we project from 2012 & 2013 emissions levels. We use 9.67 & 9.86 GtC for burning fossil fuels, gas flaring, and cement use and 0.9 GtC for land use from USDOE (2012 & 2013) CO_2 emission estimates. We assume an emissions growth of 2% per year after the year 2012 based on NOAA data. Total accumulated emissions were determined between 2012 and 2050. The IPCC determined the remaining greenhouse gas budget of 270 GtC from the year 2012 and on. Based on our projections we determine that we will blow the greenhouse gas budget in 2033. 2012 & 2013 carbon dioxide emission levels were obtained from the United States Department of Energy Carbon Dioxide Information Analysis Center (CDIAC), Available at, http://cdiac.ornl.gov. The 270 GtC greenhouse gas budget was obtained from, IPCC, WGI, AR5 2013, pp. 26–28, 103; *See also*, NOAA Reasearch, "Earth Systems Research Laboratory," Available at, http://www.esrl.noaa.gov/gmd/ccgg/trends/mlo.html.

184. Mann, Micheal E., "Earth Will Cross the Climate Danger Threshold by 2036," *Scientific American,* March 18, 2014, Available at, http://www.scientificamerican.com/article/earth-will-cross-the-climate-danger-threshold-by-2036/. Michael E. Mann, Distinguished Professor of Meteorology at Pennsylvania State University found results that are consistent with ours. His calculations, *". . . indicate that if the world continues to burn fossil fuels at the current rate, global warming will rise to two degrees Celsius by 2036, crossing a threshold that will harm human civilization."*

185. Freedman, Adam, "IPCC Report Contains 'Grave' Carbon Budget Message," *Climate Central*, October 4th, 2013, Available at, http://

www.climatecentral.org/news/ipcc-climate-change-report-contains-grave-carbon-budget-message-16569.

186. IPCC, WGI, AR5 2013, pp. 26–28, 103.

187. MIT Joint Program on the Science and Policy of Global Change, "2014 Energy and Climate Outlook," *MIT,* Available at: http://globalchange.mit.edu/research/publications/other/outlook.

188. MIT 2014 Energy and Climate Outlook; *"Fossil fuel energy continues to account for over 80% of primary energy through 2050 (despite rapid growth in renewables and nuclear), in part because the natural gas share of primary energy also increases;"* See also, International Energy Agency, World Energy Outlook 2014 Executive Summary, According to the IEA *"Policy choices and market developments that bring the share of fossil fuels in primary energy demand down to just under three-quarters in 2040 are not enough to stem the rise in energy-related carbon dioxide (CO_2) emissions, which grow by one-fifth."* Available at: http://www.iea.org/Textbase/npsum/WEO2014SUM.pdf (Check for the 2015 World Energy Outlook when available); *See also,* U.S. Energy Information Administration, "Annual Energy Outlook 2015 projections to 2040," *US EIA,* p. 15, April 2015. The U.S. Energy Information Administration predicts only 10% of the U.S energy mix to be renewable in 2040.

189. United Nations, "The Emissions Gap Report 2013," United Nations Synthesis Report, *UNEP,* 2013, Available at: http://www.unep.org/pdf/UNEPEmissionsGapReport2013.pdf.

190. Rogers, Jim, *"Lighting the World, Transforming Our Energy Future by Bringing Electricity to Everyone",* St. Martin's Press, August 25, 2015. *"...new large-scale, sustainable solutions will not only usher in a new era of light, but be an important first step in lifting people from poverty and putting them on a road of sustainable economic development. Also, a unique, transforming opportunity for Western thinkers and practitioners will be created."* See Also, The World Bank, Available at: http://www.worldbank.org/en/news/video/2014/10/14/ending-energy-poverty, *"Providing electricity and clean household fuels to the 1.2 billion who are without them, while also supporting the shift to a sustainable energy*

path are the key elements of the World Bank's approach to energy."

191. MIT 2014 Energy and Climate Outlook.

192. IPCC, "Working Group III Mitigation, AR5, Historic Trends and Driving Forces," Table 3.28a, 2014 Available at, http://www.ipcc.ch/ipccreports/tar/wg3/index.php?idp=125.

193. Meinshausen, Malte, et al., "Greenhouse-gas emissions targets for limiting global warming to 2°C," *Nature*, 458, 1158–1162, 30 April 2009, Available at, http://www.nature.com/nature/journal/v458/n7242/full/nature08017.html; *See also*, Carbon Tracker, "Unburnable Carbon, Are the world's financial markets carrying a carbon bubble," Available at, http://www.carbontracker.org/report/carbon-bubble/; *See also*, McKibbin, Bill, "Do the Math," *350.org*, Available at: http://math.350.org; *See also*, McKibben, Bill, "Global Warming's Terrifying New Math," *Rolling Stones*, July 19, 2012, Available at, http://www.rollingstone.com/politics/news/global-warmings-terrifying-new-math-20120719; *See also*, "Carbon Visuals," *350.org*, http://www.carbonvisuals.com/projects/do-the-math-supporting-a-350-dot-org-tour; *See also*, Bridger, Alan, "Note from Alan Bridger, editor-in-chief, Keep it in the Ground," *The Guardian*, Available at, http://www.theguardian.com/environment/ng-interactive/2015/mar/16/keep-it-in-the-ground-guardian-climate-change-campaign; These groups using results from the Meinshausen, Malte, et al., April 2009 study assume a more conservative remaining greenhouse gas (carbon) budget at 565 gigatons of carbon dioxide (154 GtC) than the IPCC 270 GtC remaining budget. A more conservative estimate based on a 154 GtC (565 gigatons carbon dioxide) greenhouse gas budget finds that it is 8 times greater.

194. IPCC, "Working Group III Mitigation, Historic Trends and Driving Forces," Available at, http://www.ipcc.ch/ipccreports/tar/wg3/index.php?idp=125; *See also*, International Energy Agency, "World Energy Outlook 2012," *IEA*, Nov. 2012, Available at, www.iea.org/publications/freepublications/publication/English.pdf; *See also*, McGlade, Christophe Ekins, Paul, "The geographical distribution of fossil fuels unused when limiting global warming to 2 °C," *Nature*, 517, 187–190,

January 8, 2015, Available at, http://www.nature.com/nature/journal/
v517/n7533/full/nature14016.html; *See also*, Carrington, Damian,
"Leave fossil fuels buried to prevent climate change, study urges,"
The Guardian, January 7, 2015, Available at: http://www.theguardian.
com/environment/2015/jan/07/much-worlds-fossil-fuel-reserve-must-
stay-buried-prevent-climate-change-study-says; *See also*, Vaughan,
Adam, "Earth Day: scientists say 75% of know fossil fuel reserves
must stay in ground," *The Guardian*, April 22, 2015, Available at,
http://www.theguardian.com/environment/2015/apr/22/earth-day-
scientists-warning-fossil-fuels- ; *See also*, Meinshausen, Malte, et
al., "Greenhouse-gas emissions targets for limiting global warming
to 2°C," *Nature*, April 2009; *See also*, http://www.carbonbrief.org/
blog/2015/01/meeting-climate-targets-means-80-per-cent-of-worlds-
coal-is-unburnable,-study-says/.

195.　COMMITTEE ON ENERGY AND COMMERCE, ONE HUNDRED
THIRTEENTH CONGRESS, Memorandum, September 17, 2013, *"In
December 2012, the International Energy Agency examined the global
emissions reductions that would be necessary to limit global warming
by 2°C and found that 80% of the carbon emissions allowable by 2035
are already 'locked-in' by existing infrastructure. If the world does
not take significant action to reduce emissions by 2017, then the world
will be locked in to carbon emissions levels that guarantee warming
of more than 2°C,"* Available at: http://democrats.energycommerce.
house.gov/sites/default/files/documents/Memo-EP-Climate-Change-
Obama-Administration-2013-9-17.pdf; *See also*, International Energy
Agency, *World Energy Outlook 2012*, Nov. 2012, Available at, http://
www.iea.org/publications/freepublications/publication/English.pdf.

196.　United Nations, "The Emissions Gap Report 2013, United Nations
Synthesis Report," *UNEP*, 2013, According to the 2013 UN Emissions
Gap Report, global greenhouse gas levels continue to rise rather
than decline and in fact 2010 levels were already 14% higher than
the median estimated level we must be at in 2020 to reach a least
cost pathway of meeting the 2°C target (44 $GtCO_2e$ per year), *See also*,
MIT 2014 Energy and Climate Outlook.

197.　NOAA Reasearch, "Earth Systems Research Laboratory," Available at, http://www.esrl.noaa.gov/gmd/ccgg/trends/mlo.html.

198. Nicholas Stern, Hans Joachim Schellnhuber, Jeffrey Sachs et al., "Earth Statement," *Earth League*, 2015, Available at: http://www. the-earth-league.org & http://earthstatement.org & http://www. theguardian.com/environment/2015/apr/22/earth-day-scientists-warning-fossil-fuels-; Our calculations are consistent with The Earth League findings. To determine emission levels we project from 2012 & 2013 emissions levels. We use 9.67 & 9.86 GtC for burning fossil fuels, gas flaring, and cement use and 0.9 GtC for land use from USDOE (2012 & 2013) CO_2 emission estimates. We assume an emissions growth of 2% per year after the year 2012. Total accumulated emissions were determined between 2012 to 2050. Based on our projections we determine that we will blow the greenhouse gas budget in 2033 based on the IPCC remaining 2012 greenhouse budget, 270 GtC. 2012 & 2013 carbon dioxide emission levels were obtained from the United States Department of Energy Carbon Dioxide Information Analysis Center (CDIAC), Available at, http:// cdiac.ornl.gov. The greenhouse gas budget was obtained from, IPCC, WGI, AR5 2013, pp. 26–28, 103; *See also*, "Global Carbon Project," http://www.globalcarbonproject.org/carbonbudget/15/hl-full.htm

199. *Ibid*; Our calculations show that peaking yearly emissions in 2015 allows the world to more slowly reduce emissions to a doable rate of 3% per year. If we reduce emissions by 3% per year starting in 2015 than our cumulative emissions from 2012–2050 stay below the 270 GtC greenhouse gas budget. Peaking emissions in the year 2030, results in drastic 22% per year greenhouse gas emissions reductions necessary to meet the remaining 270 GtC greenhouse gas budget between 2012 and 2050; *See also*, Paltsev, S., Reilly, J., & Sokolov, A., "What GHG Concentration Targets are Reachable in this Century," *MIT Joint Program on the Science and Policy of Global Change*, Report No. 247, July 2013; *See also*, Anderson, K. and Bows, A., "Beyond 'dangerous' climate change," 2011; *See also*, Anderson, Kevin, "Climate change going beyond dangerous," September 2012; *See Also*, Anderson, Kevin, "Duality in Climate Science," *Nature Geoscience*, Advance Online Publication, October 2015.

200. *Ibid.*

201. United Nations, "The Emissions Gap Report 2013"; *See also,* COMMITTEE ON ENERGY AND COMMERCE, ONE HUNDRED THIRTEENTH CONGRESS, Memorandum, September 17, 2013, Available at, http://democrats.energycommerce.house.gov/sites/default/files/documents/Memo-EP-Climate-Change-Obama-Administration-2013-9-17.pdf; *See also,* The Obama-Biden Plan, Office of the President-Elect, Change.Gov, Available at, http://change.gov/agenda/energy_and_environment_agenda/; *See also,* Romm, Joe, Misleading U.N. Report Confuses Media On Paris Climate Talks, Climate Progress, Nov, 3, 2015, Available at, http://thinkprogress.org/climate/2015/11/03/3718146/misleading-un-report-confuses-media-paris-climate-talks/.

202. Zoback, Mark, D. & Gorelick, Steven. M., "Earthquake triggering and large-scale geologic storage of carbon dioxide," *PNAS,* vol. 109 no. 26, pp. 10164–10168, May 2012, According to this study, *"... there is a high probability that earthquakes will be triggered by injection of large volumes of CO_2 into the brittle rocks commonly found in continental interiors. Because even small- to moderate-sized earthquakes threaten the seal integrity of CO_2 repositories, in this context, large-scale CCS is a risky, and likely unsuccessful, strategy for significantly reducing greenhouse gas emissions;"* See also, Romm, Joe, "FutureGen Dead Again: Obama Pulls Plug On 'NeverGen' Clean Coal Project," *Climate Progress,* February 5, 2015, Available at, http://thinkprogress.org/climate/2015/02/05/3619195/futuregen-clean-coal-project-dead/; *See also,* Orcutt, Mike, "Will Carbon Capture Be Ready on Time?" *MIT Technology Review,* June 29, 2012, Available at, http://www.technologyreview.com/news/428355/will-carbon-capture-be-ready-on-time/; *See also,* Hackbarth, "Carbon Capture Technology Not Ready, Says Former Energy Official," *Free Enterprise,* October 13, 2013, Available at, http://archive.freeenterprise.com/energy-environment/carbon-capture-technology-not-ready-says-former-energy-official; *See Also,* Anderson, Kevin, "Duality in Climate Science," *Nature Geoscience,* Advance Online Publication, October 2015.

203. IPCC, WGI, AR5 2013, pp. 29, 98, 469, 526, 546, 552, 575, 578, 627-629,

632-634, According to the IPCC, *"Carbon Dioxide Removal (CDR) methods are intentional large scale methods to remove atmospheric CO_2 either by managing the carbon cycle or by direct industrial processes (Table 6.14). In contrast to Solar Radiation Management (SRM) methods, CDR methods that manage the carbon cycle are* unlikely *to present an option for rapidly preventing climate change . . . The 'rebound effect' in the natural carbon cycle is* likely *to diminish the effectiveness of all the CDR methods (Figure 6.40). The level of* confidence *on the effects of both CDR and SRM methods on carbon and other biogeochemical cycles is* very low.*"*

204. Lawrence Livermore National Laboratory, "Carbon Flow Charts, USDOE," Available at, https://flowcharts.llnl.gov/commodities/carbon.

205. EIA, Today in Energy, U.S. remained world's largest producer of petroleum and natural gas hydrocarbons in 2014, Available at: http://www.eia.gov/todayinenergy/detail.cfm?id=20692.

206. Office of the Press Secretary, "President Obama Announces Historic 54.5 mpg Fuel Efficiency Standard," The White House, Available at, https://www.whitehouse.gov/the-press-office/2011/07/29/president-obama-announces-historic-545-mpg-fuel-efficiency-standard.

207. Alan Rusbridger, former editor-in-chief of Guardian News & Media from 1995–2015, started the "Keep it in the Ground" campaign, Available at, http://www.theguardian.com/environment/series/keep-it-in-the-ground. The concept of "keep it in the ground" was inspired by Bill McKibben, cofounder and senior advisor at 350.org.

208. Davenport, Coral, "U.S. Will Allow Drilling for Oil in Arctic Ocean," *New York Times*, May 11, 2015, Available at, http://www.nytimes.com/2015/05/12/us/white-house-gives-conditional-approval-for-shell-to-drill-in-arctic.html; *See also*, Keystone XL pipeline, Friends of the Earth, Available at, http://www.foe.org/projects/climate-and-energy/tar-sands/keystone-xl-pipeline.

209. Ford, Daniel, *The Cult of the Atom: The Secret Papers of the Atomic Energy Commission*, Simon & Schuster 1982 p. 30, quoting a 1945 statement reported by Dean Dietz.

Chapter 9: Nuclear Energy Spare Us the Cure

210.   *Ibid.*, p.30, 31. Dean Dietz was the science writer for Scripps Howard newspapers was the reporter.

211.   President Eisenhower's "Atoms for Peace" Speech, December 8, 1953, Before the General Assembly of the United Nations on Peaceful Uses of Atomic Energy. Available at, http://www.atomicarchive.com/Docs/Deterrence/Atomsforpeace.shtml.

212.   Sepahpour-Ulrich, "Soraya Atoms For Peace And Iran Nuclear Talks: The Theatrics Continues—From Eisenhower to Obama," *Foreign Policy Journal,* July 18, 2014, Available at, http://www.foreignpolicyjournal.com/2014/07/18/atoms-for-peace-and-iran-nuclear-talks-the-theatrics-continues-from-eisenhower-to-obama/, A historical summary of Iran's quest for advanced nuclear power.

213.   Amano, Yukiva IAEA Director General, "Statement at Opening of International Conference on Nuclear Security: Enhancing Global Efforts," *International Atomic Energy Agency,* July 1, 2013, Available at, https://www.iaea.org/newscenter/statements/statement-opening-international-conference-nuclear-security-enhancing-global.

214.   Cochran, Dr. Thomas & Christopher Paine, "PEDDLING PLUTONIUM Nuclear Energy Plan Would Make the World More Dangerous," *National Resources Defense Council,* March 2006, Available at, http://www.nrdc.org/nuclear/gnep/agnep.pdf.

215.   Makhijani, Arjun and Saleska, "The Nuclear Power Deception U.S. Nuclear Mythology from Electricity "Too Cheap to Meter" to "Inherently Safe" Reactors," *Ieer.org,* 1996, Available at, http://www.ieer.org/reports/npd.html; The authors Arjun Makhijani and Scott Saleska have expanded their work in a book of the same name published in 1999; *See also,* Makhijani, Arjun and Saleska, Scott, *"The Nuclear Power Deception: US nuclear mythology from electricity "too cheap to meter" to "inherently safe" reactors,"* Rowman & Littlefield Publishers, May 15, 1999.

216.   Biello, David, "How Nuclear Power Can Stop Global Warming,"

Scientific American, December 12, 2013, Available at, http://www. scientificamerican.com/article/how-nuclear-power-can-stop-global-warming/. A summary of their arguments is reported in the *Scientific American.*

217. D'Arrigo, Diane and Olson, Mary, "DOE's Dispersal of Radioactive Waste into Landfills and Consumer Products," *Nuclear Information and Resource Service,* May 14, 2007, Available at, http://www.clarku. edu/mtafund/prodlib/nuclear_information/Out_of_Control.pdf. A groundbreaking study examining this practice is Out of Control—On Purpose, published in 2007 by the Nuclear Information and Resource Service.

218. Makhijani, Arjun and Saleska, Scott, *The Nuclear Power Deception: US nuclear mythology from electricity "too cheap to meter" to "inherently safe" reactors,* Rowman & Littlefield Publishers, May 15, 1999. A detailed analysis of the history, lasting environment harm high costs and dangers of nuclear energy is made by Makhijani and Saleska in their 1999 book *The Nuclear Power Deception.*

219. Speech of William Von Home Jr. Executive Vice President and Chief Strategy Officer of Excelon Corp March 2, 2015 MIT Energy Conference, Available at, http://www.eenews.net/videos/1980/ transcript.

PART III: OPPORTUNITIES

Chapter 10: Hydrogen Is the Hopeful Future

220. Schiller, Mark, "Hydrogen energy storage: The Holy Grail for renewable energy grid integration," *Fuel Cells Bulletin,* Volume 2013, Issue 9, Pages 12–15, September 2013, Available at, http://www.sciencedirect.com/science/article/pii/S1464285913703328; *See also,* Geoffrey G. Holland and James J. Provenzano, *The Hydrogen Age,* Gibbs Smith, 2007.

221. President Bush 2003 State of the Union Address promotes hydrogen cars, January 28, 2003, Available at, http://electrifyingtimes.com/

state_of_the_union_2003.html.

222. Ruppert, Michael C., "Why Hydrogen is No Solution—Scientific Answers to Marketing Hype, Deception and Wishful Thinking," *From the Wilderness Publications*, 2003, Available at, http://fromthewilderness.com/free/ww3/081803_hydrogen_answers.html.

223. Lipman, Timothy, "What will Power the Hydrogen Economy? Present and future Sources of Hydrogen," Report prepared for the *Natural Resources Defense Council*, July 12, 2004.

224. Than, Ker, "One Step Closer to Artificial Photosyntheis and Solar fuels," *Caltech*, March 9, 2015, Available at, http://www.caltech.edu/news/one-step-closer-artificial-photosynthesis-and-solar-fuels-45875; *See also*, Bullis, Kevin, "A Better Way to Get Hydrogen From Water," *MIT Technology Review*, June 19. 2012.

225. Rogers, Paul, "Hydrogen highway hits roadblock," *Mercury News*, March 31, 2015, Available at, http://www.mercurynews.com/breakingnews/ci_8763807.

226. Martin, Chris, "Fuel-Cell Cars Receive Hydrogen-Station Boost From U.S.," *Bloomberg News*, April, 30, 2014, Available at, http://www.bloomberg.com/news/articles/2014-04-30/fuel-cell-cars-receive-hydrogen-station-boost-from-u-s-.

227. MODEL U CONCEPT: A MODEL FOR CHANGE, *Conceptcarz*, Available at, http://www.conceptcarz.com/vehicle/z6655/Ford_Model%20U/default.aspx. The 2003 Model U received high praise; there are practically no references to it on the internet subsequent to its display.

228. Silverstein, Ken, "Obama Administration Wants to Speed Up Hydrogen-Powered Vehicles," *IEEE Spectrum*, January 13, 2014, Available at, http://spectrum.ieee.org/energywise/transportation/efficiency/obama-administration-wants-to-speed-up-hydrogenpowered-vehicles.

229. California Energy Commission, California Investing Nearly $50 Million in Hydrogen Refueling Stations, May 1, 2014, Available

at, http://www.energy.ca.gov/releases/2014_releases/2014-05-01_ hydrogen_refueling_stations_funding_awards_nr.html.

230. Toyota, Available at, https://ssl.toyota.com/mirai/fcv.html.

231. Lane, Ben, "Hydrogen fuel cell cars," *Next Greencar*, updated, June 15, 2015, Available at, http://www.nextgreencar.com/fuelcellcars.php.

232. Green Car Reports, available at, http://www.greencarreports.com; *See also*, Voelcker, John, Nissan Leaf With 250-Mile Range: Ghosn Shows R&D Car, Video At Annual Meeting, *Green Car Reports*, June 25, 2015, Available at http://www.greencarreports.com/news/1098870_ nissan-leaf-with-250-mile-range-ghosn-shows-rd-car-video-at-an-nual-meeting; *See also*, Zach, Electric Cars 2015—Prices, Efficiency, Range, Pics, More, *EVObsession*, February 16, 2014, Available at, http:// evobsession.com/electric-cars-2014-list/.

233. Hennessey, Barbara C., "Hydrogen and CNG Fuel System Safety Research," *National Highway Traffic Safety Administration Office of Vehicle Crashworthiness Research*, January 26, 2013, Available at, http://www.sae.org/events/gim/presentations/2012/hennesseynhtsa. pdf.

234. Mearian, Lucas, "Here's why hydrogen-fueled cars aren't little Hindenburgs," *Computerworld*, November 26, 2014, Available at, http://www.computerworld.com/article/2852323/heres-why-hydro-gen-fueled-cars-arent-little-hindenburgs.html; *See also*, "Hydrogen safer than gasoline," *Enerlix, International Marketplace for Environmental Technologies*, Available at, http://www.enerlix.com/ environmental-technology/report_206.htm.

235. The History of Gasoline Retailing, "The association for conve-nience and fuel retailing," Available at, http://www.nacson-line.com/YourBusiness/FuelsReports/GasPrices_2011/ Pages/100PlusYearsGasolineRetailing.aspx. A short history of the development of the gas station.

236. European Union, "H2Aircraft-CRYOPLANE and the future of flight," Available at, http://ec.europa.eu/research/transport/projects/items/ h2aircraft___cryoplane_and_the_future_of_flight_en.htm; *See also*,

North American Renewable Energy Directory, Available at, http://nared.org/alternative-fuel-vehicle/hydrogen-powered-aircraft/. Experimental hydrogen-powered aircraft are flying today.

237. Liquid Hydrogen Fueled, "Experimental Cargo Airline," *Aircraft Engineering and Aerospace Tachnology*, Vol 51 Issue, 12, pp.17–18, 1979, Available at, http://www.emeraldinsight.com/doi/abs/10.1108/eb035584?journalCode=aeat.

238. GL Group, "Hamburg—hydrogen power for ships," *The Hydrogen Journal*, May 15, 2009, Available at, http://www.h2journal.com/displaynews.php?NewsID=156. The article describes several possible ways to configure a hydrogen- powered ship.

239. "The Hydrogen House," Available at, http://hydrogenhouseproject.org/the-hydrogen-house.html.

240. Energy storage Association, "Hydrogen Energy storage, Executive Summary," Available at, http://energystorage.org/energy-storage/technologies/hydrogen-energy-storage.

241. Hydrogen House Project, Available at, http://www.hydrogenhouse-project.org

Chapter 11: Electric Cars and Heat Pumps for Every Home

242. *Heat Pump Centre*, "Heat Pump Technology," Available at, http://web.archive.org/web/20130728151936/http://www.heatpumpcentre.org/en/aboutheatpumps/heatpumptechnology/Sidor/default.aspx, July 28, 2013.

243. *Energy.gov*, "Heat Pump Systems," Available at, http://energy.gov/energysaver/articles/heat-pump-systems.

244. *Heat Pump Centre*, "Heat Pump performance," Available at: http://web.archive.org/web/20130728151936/http://www.heatpumpcentre.org/en/aboutheatpumps/heatpumptechnology/Sidor/default.aspx, The performance of heat pumps is usually described by the Coefficient Of Performance (COP), which is the ratio of useful heat produced to the drive energy of the heat pump. The Seasonal Performance Factor

(SPF) is the average COP taken over a heating season. The Coefficient Of Performance (COP) of the residential and commercial heat pump heating systems is between 2.5–5. Fossil fuel heating systems all operate below 100%. A natural gas furnace has a maximum efficiency around 97% with COPs between 0.80–0.97. This results in heat pump efficiency of roughly 300%–500% or more for heating buildings. Some industrial heat pumps actually have efficiencies of 10,000% or more!

245. EIA, "How much energy is consumed in residential and commercial buildings in the United States?" Available at, http://www.eia.gov/tools/faqs/faq.cfm?id=86&t=1; *See also*, Kelso, D. Jordan, "Buildings Energy Data Book," *USDOE*, March 2012. Available at, http://buildingsdatabook.eere.energy.gov/; *See also*, e.g., Kelso, D. Jordan, D&R International, Ltd, "2011 Buildings Energy Data Book," *Prepared for the USDOE, PNNL*, March 2012. Available at, http://buildingsdatabook.eere.energy.gov/; *See also*, "Heating fuel choice shows electricity and natural gas roughly equal in newer homes" Available at: http://www.eia.gov/todayinenergy/detail.cfm?id=7690, *See also*, e.g., "EIA Residential Energy Consumption Survey (RECS)," Available at: http://www.eia.gov/consumption/residential/.

246. "EPA, Contact US State and Local Climate and Energy Program," Available at, http://www.epa.gov/statelocalclimate/local/topics/transportation.html; *See also,* Lawrence Livermore National Laboratory, Estimated US energy use in 2013: 97.4 Quads, Available at, https://flowcharts.llnl.gov/content/energy/energy_archive/energy_flow_2013/2013USEnergy.png; *See also,* Kelso, D. Jordan, D&R International, Ltd, "*2011* Buildings Energy Data Book," *Prepared for the USDOE, PNNL,* March 2012. Available at: http://buildingsdatabook.eere.energy.gov/.

247. EIA, "Today in Energy," Available at, http://www.eia.gov/todayinenergy/detail.cfm?id=10271&src=%E2%80%B9%20Consumption%20%20%20%20%20%20Residential%20Energy%20Consumption%20Survey%20(RECS)-b1.

248. Heat Pump Centre, "How heat pumps achieve energy savings and CO_2 emissions reduction: an introduction," Available at, http://web.

archive.org/web/20130805083932/http://www.heatpumpcentre.org/en/
aboutheatpumps/howheatpumpsachieveenergysavings/Sidor/default.
aspx; *See also*, "Heat Pump performance," Available at: http://web.
archive.org/web/20130728151936/http://www.heatpumpcentre.org/en/
aboutheatpumps/heatpumptechnology/Sidor/default.aspx; *See also*,
Goetzler, W. et al., Navigant Consulting Inc., *Ground-Source Heat
Pumps: Overview of Market Status, Barriers to Adoption, and Options
for Overcoming Barriers*, USDOE Energy Efficiency and Renewable
Energy Geothermal Technologies Program, p.49, Feb 3, 2009; *See also*,
Hutzel, W.J. & Groll, E.A., "Energy and Water Projects, Cold Climate
Heat Pump 201136," p. 5, *ESTCP Final Report 201136*, August 2013.

249.   Goetzler, W. et al, "Ground-Source Heat Pumps," *USDOE*, Feb 3,
2009; *See also*, Stevens, V. et al., "Air Source Heat Pumps in Southeast
Alaska," *CCHRC*, April 2013; *See also*, Meyer, J. et al., by Alaska Center
for Energy and Power Cold Climate Housing Research Center,
"Ground Source Heat Pumps in Cold Climates, The Current State
of the Alaska Industry, a Review of the Literature, a Preliminary
Economic Assessment, and Recommendations for Research," *A report
for the Denali Commission*, May 31, 2011; *See also*, Underland, Helge,
"On Top of the World: Arctic Air Base Warmed with Heat Pump
Technology," *GHC BULLETIN*, September 2004. *See also*, Stevens
Winter Associates, Inc. & Lis, D., "Northeast/Mid-Atlantic Air-Source
Heat Pump Market Strategies Report," *Northeast Energy Partnership
(NEEP)*, December 2013.

250.   Goetzler, W. et al, "Ground-Source Heat Pumps," *USDOE*, Feb 3,
2009; *See also*, Stevens, V. et al., "Air Source Heat Pumps in Southeast
Alaska," *CCHRC*, April 2013; *See also*, Meyer, J. et al., by Alaska Center
for Energy and Power Cold Climate Housing Research Center,
"Ground Source Heat Pumps in Cold Climates, The Current State
of the Alaska Industry, a Review of the Literature, a Preliminary
Economic Assessment, and Recommendations for Research," *A report
for the Denali Commission*, May 31, 2011; *See also*, Underland, Helge,
"On Top of the World: Arctic Air Base Warmed with Heat Pump
Technology," *GHC BULLETIN*, September 2004. *See also*, Stevens
Winter Associates, Inc. & Lis, D., *Northeast/Mid-Atlantic Air-Source*

Heat Pump Market Strategies Report, Northeast Energy Partnership (NEEP), December 2013.

251.   Goetzler, W. et al., Navigant Research Inc., "Research and Development Roadmap: Geothermal (Ground-Source) Heat Pumps," *USDOE Energy Efficiency & Renewable Energy Building Technologies Program*, October 2012. Ground & water source heat pumps are good for cold and warm environments and new high efficiency GSHP are greatly reducing their payback time. The reason they have not caught on is because, there is lack of knowhow, mass productions, the cost of the initial installation can be extremely high, and land availability & assessment can pose a challenge. The DOE states that the time to recoup the installation costs from lower utility bills is between 2 and 10 years. For new construction saving will usually be immediate. *See also*, "Choosing and Installing Geothermal Heat Pumps," Available at: http://energy.gov/energysaver/articles/choosing-and-installing-geothermal-heat-pumps.

252.   (Department of Energy), "Geothermal Heat Pumps," Available at, http://energy.gov/energysaver/articles/geothermal-heat-pumps.

253.   Hutzel, W.J. & Groll, E.A., "Energy and Water Projects, Cold Climate Heat Pump 201136," p. 5, *ESTCP Final Report 201136*, August 2013, *See Also*, Goetzler, W. et al., "Ground-Source Heat Pumps," *USDOE*, Feb 3, 2009, *See Also*, Stevens Winter, Inc., NEEP Report, Dec 2013.

254.   Safa, A. & Fung, A. et al., "Performance Assessment of a Variable Capacity Air Source Heat Pump and a Horizontal Loop Coupled Ground Source Heat Pump System, Technical Brief," *Archetype Sustainable House Series, Sustainable Technologies, Evaluation Program, 2012*, Available at, http://www.sustainabletechnologies. ca; *See also*, "Mitsubishi Cold Climate Heat Pumps Electric, Air Conditioning and Heating," Available at, http://mitsubishielectric. ca/en/hvac/zuba-central/index.html; *See also*, "Fujitsu Cold Climate Heat Pumps," Available at, http://www.fujitsugeneral.com/PDF_06/ halcyon06_brochure.pdf. Mitsubishi appears to have the greatest variety of commercialized units and offer the largest numbers of applications. Their cold-climate heat pumps can be used with the

traditional ducting systems found throughout North America. They can also use copper-piped zoned systems utilizing radiant heat for heated floors, stand-alone radiators, or individual indoor fan driven units (often termed mini-split systems). Systems with radiant floor heating are offered in Britain. Generally operating at roughly 300% efficiency, they are still able to run at 100% efficiency with temperatures as low as to 0 to -5oF and can continue to operate at 85% efficiency without backup heating at temperatures as low as -13oF.

255. Efficiency Maine, "Compare Heating Options," Available at, http://www.efficiencymaine.com/at-home/home-energy-savings-program/compare-heating-options/; *See also*, Stevens, V. et al., "Air Source Heat Pumps in Southeast Alaska," *CCHRC*, p. 42 , April 2013; *See also*, Stevens Winter Associates, Inc. & Lis, D., "Northeast/Mid-Atlantic Air-Source Heat Pump Report," *NEEP*, p. 29, Dec 2013; *See also*, Matley, R., "An alternative to oil heat for the Northeast," *RMI*, pp. 6, March 2013; *See also*, Hutzel, W.J. & Groll, E.A., "Cold Climate Heat Pump," p. 66, August 2013; *See also*, Goetzler, W. et al., "Ground-Source Heat Pumps," *USDOE*, Feb 3, 2009; *See also*, Hutzel, W.J. & Groll, E.A., Energy and Water Projects, Cold Climate Heat Pump 201136, p. 5, *ESTCP Final Report 201136*, August 2013.

256. Department of Energy, available at, http://energy.gov/energysaver/articles/geothermal-heat-pumps & http://energy.gov/sites/prod/files/guide_to_geothermal_heat_pumps.pdf; *See also,* Heat Pump Center, "How heat pumps achieve energy savings and CO_2 emissions reduction: an introduction," http://web.archive.org/web/20130805083932/http://www.heatpumpcentre.org/en/aboutheatpumps/howheatpumpsachieveenergysavings/Sidor/default.aspx; *See also*, e.g. "Heat Pump performance," Available at: http://web.archive.org/web/20130808115726/http://www.heatpumpcentre.org/en/abouthe-atpumps/heatpumpperformance/Sidor/default.aspx.

257. "TVA Energy Right Solutions," Available at, http://www.energyright.com/residential/heat_pumps.html.

258. Oak Ridge National Lab, "Energy Partners," Available at, http://web.ornl.gov/info/ornlreview/v38_1_05/article13.shtml.

259. Efficiency Maine, "Geothermal Heating and Cooling Systems," Available at, http://www.efficiencymaine.com/renewable-energy/geothermal-heating-cooling-systems/.

260. Green Mountain Power, "Vermont cold-climate-heat-pump-rental-program," Available at; http://web.archive.org/web/20140701013530/http://www.greenmountainpower.com/customers/heat-pump-rental/cold-climate-heat-pump-rental-program-/ & http://www.greenmountainpower.com/products-services/overview/heat-pump-services/.

261. Giambusso, David, "Council bill would set guidelines for use of geothermal energy," *Capital*, January 6, 2015, Available at, http://www.capitalnewyork.com/article/city-hall/2015/01/8559628/council-bill-would-set-guidelines-use-geothermal-energy.

262. "GEO Lauds New Hampshire Thermal Energy Law for GHPs," *HVACRBusiness*, Available at, http://www.hvacrbusiness.com/1287; *See also*, Runyon, Jennifer, New Hampshire Sets Thermal Renewable Energy Carve Out, *RenewableEnergyWorld.com*, June 26, 2012, Available at, http://www.renewableenergyworld.com/articles/2012/06/hew-hampshire-sets-thermal-renewable-energy-carve-out.html.

263. Custer, C., "China's government wants 5 million cars on the roads by 2020," *TechinAsia*, Available at, https://www.techinasia.com/chinas-government-5-million-electric-cars-roads-2020/.

264. Guest Contributor, "Making $1 Fuel Mainstream," *CleanTechnica*, Nov 26, 2015, Available at, http://cleantechnica.com/2014/11/26/making-1-fuel-mainstream/.

PART IV: LET'S MAKE IT HAPPEN

Chapter 13: Metropolis to Ecopolis

265. "Los Angeles Almanac," Available at, http://www.laalmanac.com/LA/index.htm.

266. "LAWDP Feed-in Tariff (FiT) Program," Available at, https://www.ladwp.com/ladwp/faces/ladwp/partners/p-gogreen/p-gg-localrenewableenergyprogram.

267. Lacey, Stephen, "Debating the Solar Federal Tax Credit: Will Expiration Kill Jobs or Make Installers Stronger?" *GreenTech Media*, March 26, 2015, Available, http://www.greentechmedia.com/articles/read/debating-the-solar-federal-tax-credit-will-expiration-kill-jobs-or-make-ins.

268. "Owens Valley Overview," Available at, http://www.ovcweb.org/OwensValley/OwensValley.html.

269. California Energy Commission "Renewable Resources Development Report," 2003, Appendix C.

270. Dora Yen-Nakafuji, "California Wind Resources," *California Energy Commission* staff White PAPER cec-500-2005-071-d, APRIL 2005; *See also*, Los Angeles Department of Water and Power, *Wikipedia*, Available at, https://en.wikipedia.org/wiki/Los_Angeles_Department_of_Water_and_Power.

271. Duke Energy, "$8-billion green energy initiative proposed for Los Angeles, News Release," September 23, 2014. Available at http://www.duke-energy.com/news/releases/2014092301.asp.

272. Linthicum, Kate, "L.A. moves to eliminate reliance on coal-powered energy," *Los Angeles Times*, March 19, 2013, Available at, http://articles.latimes.com/2013/mar/19/local/la-me-no-more-coal-20130320.

273. Power Plants Around the World, "WTE Plants in Germany—other länder," Available at, http://www.industcards.com/wte-germany.htm. Uses include warming streets, bike lanes and sidewalks in the Swedish winter.

274. John, Jeff St., "California Governor Jerry Brown Calls for 50% Renewables by 2030," *GreenTech Media*, January 5, 2015, Available at, http://www.greentechmedia.com/articles/read/calif.-gov.-jerry-brown-calls-for-50-renewables-by-2030.

275. TVA, "Appendix C—Tennessee Valley Authority/Power Distributors Energy Efficiency Initiatives," Available at, http://www.tva.gov/environment/reports/rutherford/RWD_FEIS_AppC.pdf. The program Dave started has since 1997 been called the Energy Right residential

program focusing on high-efficiency heat pumps, air conditioners, and water heaters.

276.   Computation made in Chapter 11, Freeman, S David, *Winning Our Energy Independence, An Insider Shows How,* Gibbs, Smith, 2007.

277.   "Port of Los Angeles Electric Trucks," Available at, http://portoflosangeles.org/environment/etruck.asp; *See also,* The Port of Los Angeles Regular Agenda, Available at, http://lwww.portoflosangeles.org/Board/2015/June%202015/061815_Regular_Agenda.asp. As of June 2015, the port is continuing to receive redesigned and more efficient short-haul electric trucks and remains committed on the record to advance their use.

278.   "Maglev", *Wikipedia,* Available at, https://en.wikipedia.org/wiki/Maglev.

Epilogue: The Path Ahead—"A Call to Action"
279.   Freeman, S. David, *Energy: The New Era*, Vintage Books, A Division of Random House, Inc. New York pp. 50-51, 1974.

INDEX

Made in the USA
Charleston, SC
31 December 2016